THE GRAVITY THEORY OF MASS EXTINCTION

THE GRAVITY THEORY OF MASS EXTINCTION

A new unified theory of mass extinction explains the rise and fall of the dinosaurs.

Third Edition

By John Stojanowski

Pangea Publications, LLC
Staten Island, New York

ISBN: 978-0-9819221-4-0

First Edition- published July 2008
Second Edition-published April 2011
Third Edition-published January 2012

Graphics created with CoPlot

10 9 8 7 6 5 4 3

In memory of my parents,
Stanley and Sophie,
from humble beginnings,
constrained by the depths of the Great Depression
and its dark aftermath,
from whom I learned that hard work is the only
reliable leveler of the playing field of life.

To doubt everything or to believe everything are two equally convenient solutions; both dispense with the need for thought.

Henri Poincare

CONTENTS

Preface   ix

Introduction   xi

PART I: MASS EXTINCTIONS

Chapter 1:   The Big Five Mass Extinctions   1

PART I I: THE ASTEROID IMPACT THEORY OF EXTINCTION

Chapter 2:   The Clay Layer at Gubbio   7

Chapter 3:   Birth of the Asteroid Impact Theory   11

Chapter 4:   The Rumbling of Dissent   23

PART III: THE VOLCANIC THEORY OF EXTINCTION

Chapter 5:   The Opposition Gathers   27

Chapter 6:   The Ensuing Debate   31

PART IV: THE GRAVITY THEORY OF MASS EXTINCTION

Chapter 7: Overview   35

Chapter 8: The Birth of The Theory   41

Chapter 9: The Evolution of The Theory   51

Chapter 10: The Eureka Moment of the Theory   75

Chapter 11: The P-T Extinction- A New Explanation   85

Chapter 12: The K-T Extinctions
    Terrestrial Extinctions   99
    Marine Extinctions   108

Chapter 13: The Triassic-Jurassic Extinctions   137

PART V: DENOUEMENT

Chapter 14: The Earth's Core And Surface Gravity   141

Chapter 15: Sea-Level Changes And Surface Gravity   151

Chapter 16: Summation   157

PART VI: REENFORCEMENT

Chapter 17: The Pangea A vs. Pangea B Controversy
    Supports The Gravity Theory Of Mass Extinction   141

Chapter 18: Physics Of The GTME   151

Chapter 19: Loose Ends   157

Chapter 20: Confirmation Of The GTME   192

References   202

Index   205

PREFACE

In 1911, a German scientist by the name of Alfred Wegener developed a theory about the Earth's continents. It was not only the apparent fitting together of the coastlines of the continents bordering the Atlantic Ocean as though they were part of a global jigsaw puzzle that drew his attention. His study of the fossils on those continents, and the similarity of contemporaneous species that formed those fossils convinced him that the continents were once joined together. He also noted that when the continents of Africa and South America are fitted together, the mountain ranges and coal deposits were continuous and uninterrupted.

In 1915 he expressed those ideas in his book *The Origin of Continents and Oceans*. The response to his theory of continental drift was met with invective by his fellow scientists:

"If we believe this hypothesis, we must forget everything we have learned in the last 70 years and start all over gain."

and,

"Anyone who valued his reputation for scientific sanity would never dare support such a theory."

The idea that the terra firma was not a fixed, unmovable outer shell of the Earth was considered pure heresy. Wegener's inability to come up with a mechanism for the movement of the continents was the biggest stumbling block for acceptance of his theory. At the time, the thought of continents grinding over what was thought to be a solid hard subsurface seemed counterintuitive.

His theory was put on a shelf for the following few decades. It wasn't until the 1950s when the discoveries related to paleomagnetism revived his theory. When it was realized that the bilateral upwelling of magma from the mid-ocean ridges would result in the movement of the underlying tectonic plates as though they were on a giant conveyer belt, the Continental Drift Theory (now known as the Theory of Plate Tectonics) gained widespread acceptance. Unfortunately, Wegener was not around to savor that moment.

The theory that I developed, the Gravity Theory of Mass Extinction (GTME), may follow the same path that Wegener's theory did. Many people will find it very hard to conceive of fluctuations of the pull of gravity at the Earth's surface. They were able to see the effects of gravity on the Apollo astronauts when they walked on the Moon but the thought that some time in the distant past, if they could be transported back to that time, they would have weighed significantly less than they do today. This is something that is inconceivable to them.

The basics of the theory explained here have been available on the Internet since 2004 on a website I developed. By publishing this book, which has expanded the original theory to a unified theory of mass extinction, I hope to expose this theory to a broader audience.

The 2nd edition introduced powerful evidence, based on paleomagnetism, that the Earth's core elements (inner/outer core & dense lower mantle, which make up about 85% of the earth's mass) moved away from their current geocentric position. This evidence is derived from the Pangea A vs. Pangea B controversy which has split paleomagnetists into two groups. The GTME reconciles the two factions and explains why the Pangea A model is the valid one. Serendipitously, paleomagnetism not only came to the rescue of the Continental Drift Theory but the Gravity Theory of Mass Extinction as well.

Part I through Part V of the 1st edition are repeated, unchanged. They describe how this theory evolved from one with a faulty foundation, i.e., that the proposed gravitational changes were due solely to the continental consolidation of Pangea. The theory evolved to include the movement of the core elements, which is the keystone of the theory. Part VI introduces and explains the paleomagnetic evidence which, in my opinion, authenticates the GTME.

This 3rd edition adds *Chapter 20* which is based upon a recent hypothesis by French scientists that the Earth's continental tectonic plates control geomagnetic field reversals. My theory is based on the linkage between the core elements and continental plates; the cited hypothesis provides overwhelming confirmatory support for this theory.

INTRODUCTION

Prior to the 1980s, dinosaur extinction was a subject that received only modest attention from the paleontological community. After all, it was believed that dinosaurs, like any other living species, would prosper for some indeterminate period and then fade away. Scientists refer to this process with the an odd description: "racial senescence." This was the observed cycle of life for all fauna and flora in the annals of the Earth.

The pronouncement, around 1980, that the dinosaurs and other animals were the victims of the impact of an extraterrestrial object sent shock waves through the paleontological community in much the same way that the purported "space rock" had done to the Earth's surface some 65 million years ago. The resultant stratification of paleontologists, geologists and other science professionals created two warring camps that are still at odds today. Among those who opposed this theory, proposed by the father-son team of Luis and Walter Alvarez, were well respected members of the scientific hierarchy. The opposition coalesced around an alternate explanation for the extinctions, the Volcanism Theory of Extinction. Today, there is still a split within the scientific community between those that support the impact theory and those that support the volcanism theory. So much attention has been focused on these two theories that any possibility of an alternate theory entering the limelight has been reduced to near zero.

This book describes a new theory of extinction published in 2004 through 2012 and was initially introduced on the Internet in 2004. This theory, written by myself, provides an alternate to the bolide impact and volcanism theories just mentioned. The proposed theory, The Gravity Theory of Mass Extinction (GTME), is applicable to all of the major mass extinctions. However, most of the emphasis will be directed to the Cretaceous-Tertiary (K-T) extinctions. The basic concept, that tectonically driven gravitational changes cause extinction, can best be demonstrated using the K-T extinctions because they have been researched the most.

The Gravity Theory of Mass Extinction attributes the K-T extinctions, and others, to a gradually changing gravitational field at Pangea's surface. This gravitational change is linked to the movement of the Earth's tectonic plates and core elements (i.e., inner and outer iron cores & lower mantle) and is an ongoing process that continues even today. In addition, the extreme gigantism of the dinosaurs and other life forms is attributed to the lowered surface gravitational field during the Mesozoic Era. The causative role of core element movement in the eruption of the massive flood basalt volcanism at the Permian-Triassic boundary (known as the Siberian Traps) and at the Cretaceous-Tertiary boundary (known as the Deccan Traps) are explained by the GTME.

As explained in the Part VI of this book, the most powerful support of this theory comes from a controversy that has arisen in the science of paleomagnetism, the Pangea A vs. Pangea B debate. Paleomagnetists, when reconstructing the position of continents in the distant past, rely on the Geocentric Axial Dipole (GAD) model of the Earth's magnetic field. When they examined the period when Pangea existed, they encountered a stumbling block; they had major problems using GAD to support the widely accepted Pangea A model. An alternate Pangea B model was introduced to supplement the other model although the Pangea B model has been accepted by only a small minority of paleomagnetists. The non-GAD model posited by the GTME is able to resolve the discrepancy between the two models and in doing so proves that the Earth's core elements did move away from their current geocentric position, and therefore, based on the laws of physics, resulted in a change in surface gravitation on Pangea. As Pangea broke apart and its constituent land masses moved latitudinally, surface gravity began to increase. This rate of increase was very high at the end of the Mesozoic Era coinciding with high extinction rates. *Chapter 18, The Physics of GTME* explains this.

Finally, there is the possibility that the asteroid (or comet), the purported "Dinosaur Killer" that struck Chicxulub, Mexico about 65 mya, had its trajectory distorted by the gravitationally induced effect of the Earth's wobble in the Cretaceous Period, which has been reported in scientific publications, thereby sending it on its deadly rendezvous with planet Earth. Since scientists cannot offer an acceptable cause for the wobble but the GTME can, the following conclusion can be drawn:

The uncanny coincidence of the temporal relationship between the K-T extinctions, the Deccan Traps volcanism and the bolide impact are the result of movement of the Earth's core elements, reacting to movement of the Earth's continental plates. It also follows that since continental configurations have varied widely prior to the Mesozoic Era, other mass extinctions may well have been caused by gravitational changes at the surface of the Earth.

PART I: MASS EXTINCTIONS

CHAPTER 1: THE BIG FIVE MASS EXTINCTIONS

Five extinction events comprise what are commonly referred to as the "Big Five" mass extinctions. The following chart displays the relative magnitude of the extinctions.

Number of
Marine Genera:

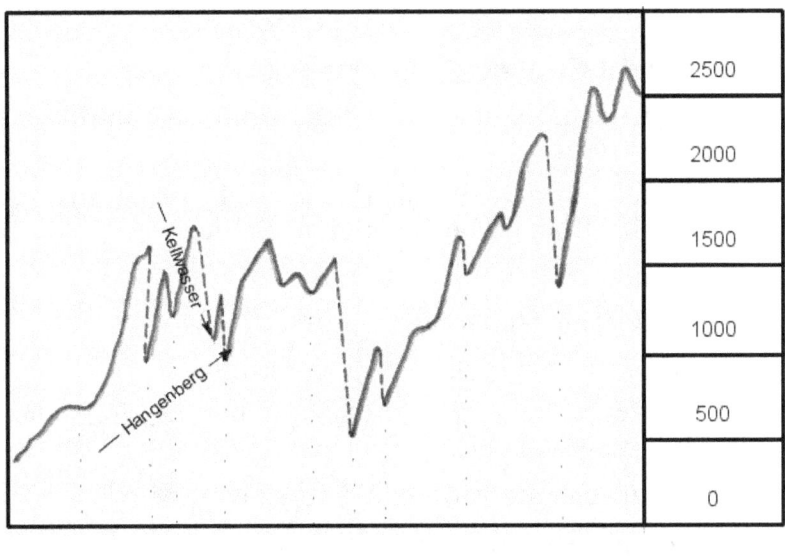

Camb.Ord/Sil.Dev/Carb P/T/Juras.   Cret/Ter.

FIG. 1-1 Marine Genera-Big 5 Extinctions

There is no precise definition of what a mass extinction is. There is a continuous background extinction of species that occurs,

1

MASS EXTINCTIONS

usually at a local level. However, to rise to the level of a mass extinction, an extinction must be eustatic (i.e., global) in its effect on life. Terrestrial, both flora and fauna, and marine life must be negatively affected. The number of species eliminated must reach a certain threshold; 50% seems to be an accepted level. Finally, the extinction interval must be in a narrow geological timespan. This is generally within a few million years although the late-Devonian mass extinction extended close to 20 million years.

Although the current period is considered by many to be the "Sixth Extinction", there are five extinction periods that are recognized as mass extinctions and they are appropriately referred to as the "Big Five."

The five events, or more accurately periods, are:

1. **Ordovician-Silurian** (~445 mya)

- Most advanced life-forms at this time were marine, not terrestrial.
- Sea-level drop (regression) followed by rise (transgression) possibly from glacier formation/melting.
- 100 marine families disappeared.
- 25% marine families disappeared.
- 50-60% marine genera disappeared.
- Many conodonts, tribolites, graptolites, brachiopods and bryozoans disappear.
- Cooling climate often cited but not convincing.

2. **Late Devonian** (a/k/a Frasnian-Famennian) (~364 mya)

- 22% marine families disappeared.
- 57% marine genera disappeared.
- 75% of all species disappeared.
- Many survivors were deep water or high latitude relatives.

- Severe extinction for benthic (bottom dwelling) marine organisms in shallow water.

- Major extinction of reef builders.

- Primarily a marine extinction including ammonites, acritarchs (microfossils), placoderms (armored, jawed fish).

- Plants increased in size from 30 cm to 30 m.

- Some members of Agnathan (jawless fish with spiking plates) suddenly disappear.

- First tetrapods appear.

- Extinction believed to be of long duration (20-25 my).

- Two main events recognized:
  Kellwasser Event (~378 mya)
  Hangenberg Event (~360 mya)

- Sea-level regression occurred.

- Cooling climate often cited but not convincing.

3. **Permian-Triassic** (~251 mya)

- 95% all species disappeared.

- 53% marine families disappeared.

- 84% marine genera disappeared.

- 70% land species (incl. plants, insects, vertebrates) disappeared.

- Pangea in final stages of consolidation.

- Massive flood basalt volcanism (Siberian Traps) occurred at this time but may post-date extinctions.

- Regression previously thought to occur at boundary; transgression now accepted.

MASS EXTINCTIONS

4. **End-Triassic** (a/k/a Triassic-Jurassic) (~214-199 mya)

- 22% marine families disappear.
- 52% marine genera disappear.
- CAMP (Central Atlantic Magmatic Province) flood basalt volcanism occurred but is believed to post-date extinctions.
- Breakup of Pangea begins - Atlantic Ocean begins to form.
- Large amphibians disappear along with conodonts.
- Mammal-like reptiles (therapsids) disappear allowing rise of dinosaurs.
- Ammonites almost disappear but a single lineage survives and radiates in Jurassic.
- 70 terrestrial families disappeared.
- Two views on timing and duration of extinctions:
  1. Short duration of about 10,000 years at boundary.
  2. Multiple prolonged events in Norian (~216 mya) and Rhaetian (~203 mya) Stages of Triassic Period.

5. **End-Cretaceous** (a/k/a Cretaceous-Tertiary) (~65 mya)

- 16% marine families disappear.
- 47% marine genera disappear.
- 18% land vertebrate families disappear.
- Dinosaurs fade over extended period before boundary and then disappear completely.
- Large marine reptiles disappear, ichthyosaurs long before, mosasaurs and elasmosaurs closer to boundary.
- Ammonites fade in pulses throughout Cretaceous and disappear completely in early Danian Stage after boundary.
- Many microscopic marine animals such as planktic foraminifera and coccoliths experience heavy losses at boundary.

- Massive flood basalt volcanism (Deccan Traps) occur before, during and after extinctions.

- Bolide impact (at Chicxulub) occurs near boundary.

The Gravity Theory of Mass Extinction (GTME), described in this book will focus primarily on mass extinctions that have occurred in the last 250 million years; the Permian-Triassic, End-Triassic and primarily the End-Cretaceous. This should not be interpreted as equivocation concerning the causes of the prior extinctions. A much greater body of information concerning the later extinctions has been accumulated and a more accurate analysis of the causes of those extinctions can be made.

Briefly, the most recognized causes of mass extinction can be summarized as:

Volcanism...the altering of the atmosphere with cooling agents (e.g., sulfur compounds); warming agents (e.g., carbon dioxide, methane); poisoning agents (e.g.,hydrogen sulfide). The effects on bodies of water would be the same as acid rain contamination.

Bolide Impact ...the effects would be similar to that of volcanism but would occur in a much shorter time interval.

Global Warming...could be a result of volcanism and bolide impact. Might be caused by consolidation of land masses (e.g., Pangea) and changes in sea-level (i.e., alteration of Earth's albedo).

Regression... (i.e., the lowering) of sea levels could eliminate shallow marine habitats.

Transgression... (i.e., the rising) of sea levels could bring anoxic (i.e., low oxygen) water into shallow coastal areas.

Glaciation.. cooling eliminates habitats.

Most contemporary publications attribute mass extinctions to one or more of the above described causes and the majority will invoke either bolide impact or volcanism. Recently, in *ScienceDaily* (10/25/06), the Press/Pulse Theory of Mass Extinction was the subject of one of the articles. It rejects the single "all-or-nothing" cause for mass extinctions. The Press/Pulse Theory posits long term extinction pressures coincident with sudden catastrophic extinction pulses.

Press/Pulse is supported by paleontology professor Nan Crystal Arens and 2006 graduate Ian West of Hobart and William Smith Colleges in Geneva, NY, USA. Their study considered "Press" to be times of massive volcanic eruptions and "Pulse" to be times of impact events. The results of the study, according to the article, are that elevated extinction levels occur only when the "Press"and "Pulse' forces coincide.

It is interesting that the Gravity Theory of Mass Extinction (GTME) supports the concept of Press/Pulse but not with volcanism/impact components. The GTME posits the changing of (the Earth's) surface gravitation as the "Press" and, when addressing certain marine extinctions, sea-level transgression as the "Pulse." Or, to be more precise, "Pulses" are rapid accelerations in surface gravity which cause regressive/transgressive sea-level couplets (i.e., rapid lowering and rising of sea levels). This concept will be described later in this book.

PART II: THE ASTEROID IMPACT THEORY OF EXTINCTION

CHAPTER 2: THE CLAY AT GUBBIO

Even though the words "hypothesis" and "theory" have very different meaning to those who study science, the latter will be used exclusively in this book to describe a scientific explanation for some phenomenon. Just as we refer to the "The Flat Earth Theory" much more than "The Flat Earth Hypothesis" without regard to its acceptance or validity, the same will apply here. In addition, the use of "mya" (million years ago) will be used in lieu of the more recent "my."

During the late 1970s geologist Walter Alvarez, of the University of California at Berkeley, became interested in the lithified formations located in northern Italy at Gubbio. Those formations were originally formed at the bottom of the sea during the Mesozoic Era and earliest Cenozoic Era and were uplifted during the formation of the mountains in the Appennine range. The raised rock formation there are a pink limestone known as the Scaglia rossa (which means red scaly rock). The pink color has been attributed to the presence of the iron oxide mineral hematite. There were several narrow clay layers in this formation and it was known that one of these layers seemed to be at the transition between the Cretaceous and the Tertiary (K-T) Periods. This boundary is sometimes referred to as the K-P (Cretaceous/Paleogene).

Apparently, Walter Alvarez and his colleagues were trying to accurately date these limestone and clay layers by gathering data on the magnetostratigraphic time stamp embedded in them. This magnetic fingerprint is established when material, which contains microscopic magnetic material, is solidified (such as lava or sedimentation). The magnetic direction or polarity is determined by

7

whether the Earth's magnetic north and south poles were normal (as they are today) or reversed (i.e., the North and South pole switched position). The magnetic time stamp, because of the long duration of the "chron" (i.e., period of same magnetic polarity), can be from a few hundred thousand years to millions of years. Therefore, any dating based on magnetostratigraphy alone would not be very accurate. He returned to the United States with rock samples of the now well known clay layer deposited around 65 mya along with samples from the overlying and underlying rock. His intention, apparently, was to determine the length of time the one centimeter thick clay layer, which appeared to be devoid of fossils, had taken to form.

Walter Alvarez discussed this problem with his father Luis. Luis Alvarez was a Nobel Prize recipient in physics, having worked extensively in nuclear physics including the development of the atomic bomb. He rode in an aircraft that accompanied the Enola Gay when it dropped the first atomic bomb on Hiroshima. Luis knew that the quantity of rare earth element iridium in the rock samples could be tested. He reasoned that since the tiny meteorites that continually bombard the Earth's atmosphere deposit a fairly low level, continuous blanket of iridium, the level in the clay layer could be compared to the other rocks where the sedimentation rate was accurately measured. The figures from the other rocks would then be a standard for comparison to the clay layer. If the iridium level was higher in the clay layer than that of the standard, it would indicate that the sedimentation rate of the clay was slower than that for the other rocks. The ratio of the two could then be used to determine the time it took to form the clay layer.

It turned out that the iridium levels in the clay layer sample were far higher than could be explained by a slower sedimentation rate. The deposition time would have been over a million years for the one centimeter clay layer. The Alvarezes then performed the same test on the contemporaneous rocks of the clay layer at Stevns Klint, Denmark and found that the iridium level was even higher than at Gubbio. After performing a similar test at a location in New Zealand and finding elevated iridium levels, they concluded that the iridium anomaly was a global phenomenon and sought to find a cause for the iridium spike.

Initially, they thought that a supernova, the collapse of a nearby star, could have showered the Earth with the iridium. After searching for one of the telltale signs of a supernova, an isotope of the element plutonium, and not detecting it, the supernova theory was abandoned.

Chris McKee, an astronomer at Berkeley, suggested that an asteroid could have caused the iridium spike. Initially, the Alvarezes were skeptical, just as many of the proponents of the Volcanism Theory of Extinction are today, of whether a single bolide impact could result in a worldwide distribution of the iridium. It has to be noted that once an asteroid enters the Earth's atmosphere and takes on the glow from friction with the atmosphere, it is called a meteor. If it is able to eventually survive to make contact with the ground, it is called a meteorite.

What convinced the Alvarezes to pursue McKee's asteroid suggestion was the research of the 1883 Krakatoa volcanic eruption near Sumatra. The explosive eruption destroyed the island it was on and the ejecta from the volcano caused spectacular red sunsets in many parts of the world for several years. They opined that if the Krakatoa eruption could send enough material into the stratosphere where it lingered for two years, a large enough asteroid could produce a much more detrimental effect by blocking the sun and severely restricting photosynthesis. With that, the concept of a nuclear winter had been introduced into the K-T extinction debate.

CHAPTER 3: BIRTH OF THE ASTEROID IMPACT THEORY

The Alvarezes were now ready to formally introduce the impact theory to explain the K-T extinctions. Along with their colleagues Frank Asaro and Helen V. Michel, the Alvarezes submitted the theory entitled:

"Extraterrestrial Cause For The Cretaceous-Tertiary Extinction" to the journal *Science*. The 13 page paper appeared in the June 6, 1980 issue (Volume 208, Number 4448).

The Science paper starts off describing the perceived extinction near the K-T transition. It even mentions one of the facts about the extinctions which critics of the impact theory still cite:

"On the other hand, some groups were little affected, including the land plants, crocodiles, snakes, mammals, and many kinds of invertebrates."

No rationalization of this apparent contradiction of the impact theory is explained. How could cold-blooded reptiles, which would be extremely sensitive to a prolonged drop in temperature for several years, according to the theory, survive? How could the same reptiles and amphibians survive the acid rain which is believed to accompany a massive impact? The paper doesn't mention frogs, turtles, salamanders or birds, to which the same questions would apply. Plants are casually mentioned but the paper doesn't mention a prominent anomaly. Tropical plants, which are very sensitive to cold temperatures fared much better than temperate plants. There didn't seem to be an extensive analysis of the fossil record before the paper was published. It seems like there was a rush to be the first to establish a relationship between a bolide impact and extinction and the hope that subsequent research would support that claim.

THE ASTEROID IMPACT THEORY OF EXTINCTION

The study of the clay layer at Gubbio in northern Italy, previously mentioned, is the focus of the research entailed in the report. The presence of microfossils in the surrounding limestone is described. The presence of the foraminifera *Globotruncana* below the clay layer and its absence above the clay layer is noted. Planktonic foraminifera are unicellular algae with a shell consisting of calcareous scales.

Also noted is the abundance of the smaller foraminifera *Globigerina eugubina* above the clay layer. This species is much smaller than the foraminifera *Globotruncana*.

No opinion is given as to the reason for the faunal change just mentioned. Can it be assumed that the "Lilliput Effect" is the implied explanation? The "Lilliput Effect" is the belief that living organisms that are smaller have a much greater chance of survival during an environmental crisis. A different study concerning the size reduction of bivalves across the K-T transition came to the following conclusion:

> **"The protracted nature of the size decrease after the K/T and the fact that the subgenera responsible do not appear to be opportunists or ecological generalists makes it unlikely that the patterns documented in this study are an example of the Lilliput Effect."**

Could the same be true for the forams? The word "forams" is a common abbreviated reference to foraminifera.

The Gravity Theory of Mass Extinction explains the above size differential on the basis of an increasing gravitational field whereby any marine animal that has a calcareous appendage (e.g., a shell) is more likely to be affected by an increasing gravitational field due to the altered buoyancy.

Another micro-marine organism affected at the K-T transition is the coccoliths, as noted in the *Science* report:

> **"The coccoliths show an abrupt change with the disappearance of Cretaceous forms, at exactly the same level as the foraminiferal change....."**

Coccoliths are unicellular algae with a shell consisting of minuscule **calcareous** scales.

The following is a drawing of the boundary based on the descriptive text on the second page of the *Science* report.

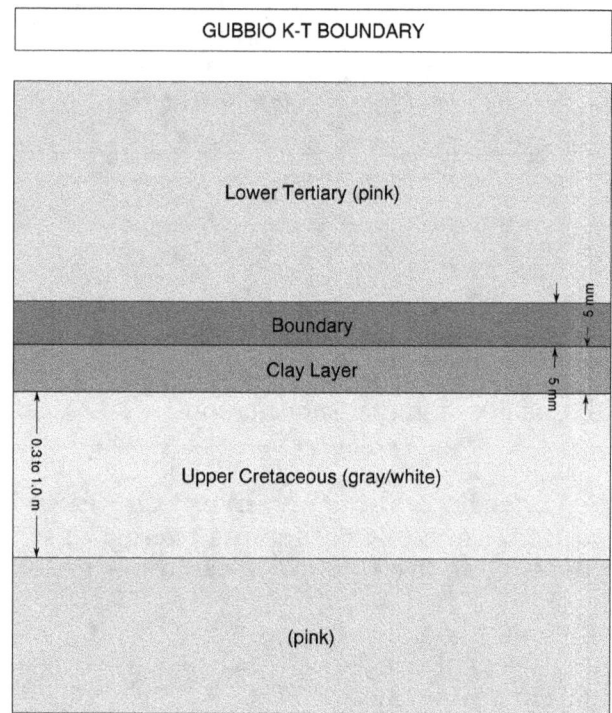

FIG. 2-1 K-T Boundary at Gubbio

The description of the rock layers, as shown in the above figure, is interesting. The gray/white section of the limestone below the clay layer stands out. See the photo on page 17. No explanation for this anomaly is proffered yet it is visually one of the most striking characteristics of the boundary formation. The paper notes the gradational change in color from pink to gray/white in the upper Cretaceous (not apparent in above diagram) and an abrupt change in color in the central part of the clay layer. At Contessa, the bottom 5 mm of the clay layer is gray and the upper 5 mm is red in color.

The report highlights the observation that the near extinction of both calcareous planktonic foraminifera and calcareous nanoplankton occurred at about the same time. Analysis of terrestrial extinctions are minimally mentioned.

There were several clay layers in the 155 million year span of the Appennine range. Aside from the K-T clay layer, no explanation for the formation of these other clay layers was offered. If the others also have a worldwide presence, then the bolide impact explanation for the K-T clay layer has to be questioned. The iridium presence in the K-T clay, whether extraterrestrial or not, would then be purely coincidental. The Appennine formation was formed under pelagic conditions. In other words, this formation was initially located in a deep sea or ocean when formed so that local disturbances most likely did not influence sedimentation or produce anomalies. It would be interesting to know if any of the other clay layers correspond to periods of extinction.

THE ASTEROID IMPACT THEORY OF EXTINCTION

Any criticism in this book of the Alvarez Impact Theory of Extinction is not directed toward the authors of that theory. Many writers have directed their opposition to that theory in a personal way, which is unfortunate. The Alvarezes and their colleagues are responsible for the vast scientific scholarship in the field of ancient extinction which was initiated as a direct result of their 1980 publication. This book, as well as hundreds of others, would not have been written if attention had not been focused on the K-T extinctions.

They must also be given credit for the current efforts to track asteroids that may one day pose a threat to planet Earth. In 2004, a 1300 foot wide asteroid named Apophis was discovered. In the year 2029, the asteroid is expected to skim by the Earth at a distance which is closer than that of some communication satellites. Although the probability of a massive asteroid colliding with the Earth is extremely small, the possibility cannot be ignored. I believe that the Alvarez Impact Theory of Extinction can be given credit for raising the awareness of the danger from near-Earth objects.

Having made the above point, I'll continue with the critique of the impact theory.

Below is a photo of the Alvarezes at the boundary clay layer at Gubbio.

Photo of Luis Alvarez and Walter Alvarez at Gubbio

Photo courtesy Univ. of California Lawrence Berkeley National Laboratory

Since the clay layer and surrounding limestone were in a deep sea when formed, the striking change in the limestone color below the clay layer indicates that some major change must have been taking place for a long time before the clay layer was deposited. Therefore, it would not have been caused by the Chicxulub impact.

THE ASTEROID IMPACT THEORY OF EXTINCTION

The measurement of the iridium levels at Gubbio and Stevns Klint, Denmark are described next in the report. The following is an adaptation of the part of the iridium graph in the report that is of interest. Note that the graph has a somewhat saw-tooth pattern with an almost vertical rise at the clay layer and a slow exponential decay rate above the clay layer.

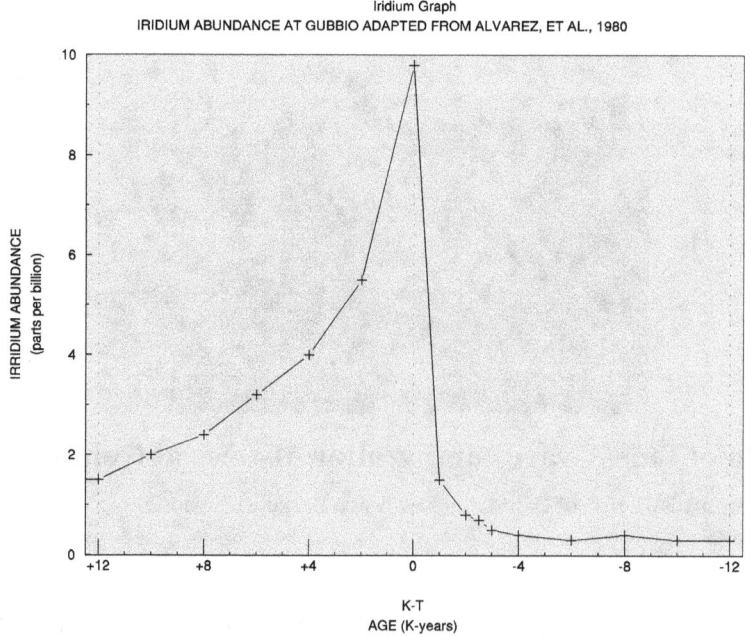

FIG. 2-2 Iridium Spike according to Alvarez, et al.

This is the iridium pattern one might expect from the impact of an iridium bearing asteroid. However, in 1987-1990, other

18

investigators went to Gubbio and Stevns Klint to perform the iridium analysis and their results were quite different. Below is a representation of the graph from the research of Robert Rocchia, of the Centre des Faibles Radioactivites in France, of the iridium layer at Bottacione, Italy. There is an elevated level of iridium for a considerable time before the clay layer was deposited. Another study from Crocket, et al., displays a similar result.

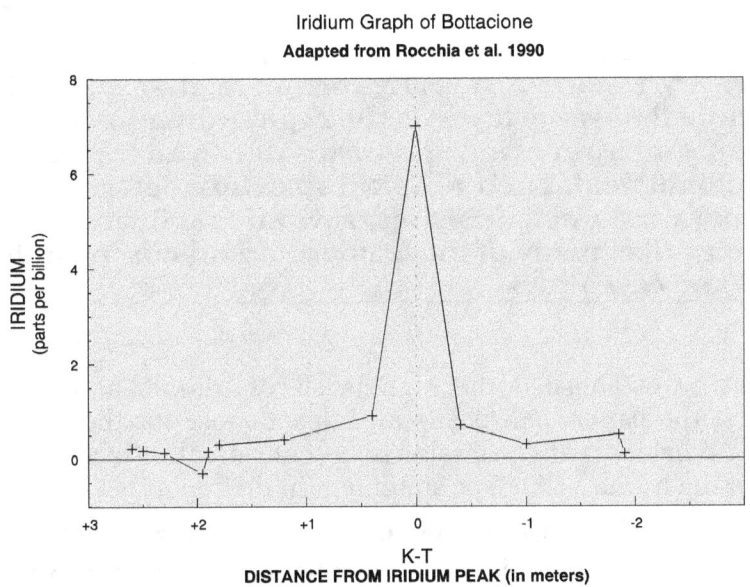

FIG. 2-3 Iridum Peak according to Rocchia, et al.

19

THE ASTEROID IMPACT THEORY OF EXTINCTION

One question that has to be asked is: Why is there a major discrepancy between the iridium graph in the 1980 *Science* journal and that of later investigators, such as Rocchia, of the same location? What is interesting is that the 1990 Rocchia et al., graph indicates a rapid rise in iridium levels between 1.5 to 2 meters below the peak, which seems to correspond to the transition from the pink limestone to the gray/white limestone described earlier. Certainly this would open the possibility of a different explanation, perhaps a volcanic one, of the source of the iridium at the K-T transition.

> **The Gravity Theory of Mass Extinction might explain the gray/white limestone layer as the rapid extinction of calcareous micro-marine organisms due to an increasing gravitational field. Their physical structure determines their buoyancy. An increasing gravitational field would adversely affect many of those microorganisms including forams and coccoliths.**

Having established the high levels of iridium at the K-T boundary, the paper then posits an impact cause for the iridium anomaly. What is glaringly missing is any consideration of a volcanic source of the iridium (Ir). One statement in the report is:

> **"Rocks from the upper mantle (which has more Ir than the crust) have less than 20 ppb (parts per billion) and are therefore an unlikely worldwide source."**

Perhaps it wasn't known in 1980 that iridium was present in ejecta from "hot spot" volcanoes such as those that were, and still are, forming the Hawaiian Island chain. However, it was well known that the heavier elements coalesced toward the Earth's center billions of years ago and that iridium would most likely occur in much higher concentrations deeper in the Earth. Again, the authors of the report seem to be intentionally stifling any consideration of a volcanic explanation for the high iridium levels in the clay layer.

The next few pages of the report are devoted to a detailed analysis of whether the iridium spike could have come from a supernova. That section is entitled "Negative Results of Tests for the Supernova Hypothesis." Again, the depth to which the supernova theory is explored and the omission of a volcanic explanation must be highlighted.

The balance of the report describes the authors' support for the impact theory of extinction and is summarized with the following points:

- The mean time to collision between an asteroid greater than 10 km in diameter and the Earth is estimated to be every 100 million years. It must be pointed out that this estimate was based on the apparently inaccurate assumption that earlier extinctions were the result of bolide impact.

- Airborne debris from the 1883 Krakatoa volcanic eruption took 2.0 to 2.5 years to completely disappear, therefore the comparable fallout from the K-T asteroid impact would be "in a few years."

- The energy of the impacting asteroid would be equivalent to 10^8 megatons (100,000,000,000,000 tons) of TNT.

THE ASTEROID IMPACT THEORY OF EXTINCTION

BIOLOGICAL EFFECTS (according to Alvarez, et al.**)**

- The disruption of photosynthesis (for several years) would deplete the food chain and lead to the marine extinctions (e.g., the belemnites, ammonites and marine reptiles).

  As will be shown in a subsequent chapter, the ammonites went extinct in phases and survived beyond the K-T boundary.

- Land plants would die or stop producing new growth but would regenerate from seeds, spores and root growth when darkness ended.

- Large herbivorous and carnivorous animals would be eliminated and the smaller animals would survive on decaying vegetation and insects.

- It is noted that among bottom dwelling marine invertebrates, some went extinct and some survived.

  The report follows with the statement "we will not go further into this matter....." No followup on this observation was made. No attempt is made to account for this important anomaly of the K-T boundary extinctions.

CHAPTER 4: THE RUMBLING OF DISSENT

A lot of research has been done on the K-T extinctions since the 1980 report in the journal *Science* was published. Some of that research supports the impact theory, but there have been more data published which now casts a large shadow of doubt about the viability of that theory.

The theory proposed in 1980 has become known as the Alvarez Impact Theory and it generated immediate controversy. Paleontologists who had studied the late Cretaceous fauna and had believed in a gradual extinction were quick to discount the Alvarez theory.

Dr. Robert Bakker, the well known paleontologist, famous for his writing on dinosaur morphology and theory that dinosaurs were warm-blooded, had this to say:

> **"The arrogance of those people is simply unbelievable. They know next to nothing about how real animals evolve, live and become extinct. But, despite their ignorance, the geochemists felt that all you have to do is crank up some fancy machine and you've revolutionized science. The real reasons for the dinosaur extinctions have to do with temperature and sea-level changes, the spread of diseases by migration and other complex events. In effect, they're saying this: 'We hi-tech people have all the answers, and you paleontologists are just primitive rock hounds.'"**

The discourse between the opposing factions was not what would be expected of scientists. Luis Alvarez is known to have said of the paleontologists who opposed the impact theory:

THE ASTEROID IMPACT THEORY OF EXTINCTION

"I don't want to say bad things about paleontologists, but they're not very good scientists, they're more like stamp collectors."

Another paleontologist, Dr. William A. Clemens of the University of California at Berkeley, had found dinosaur fossils at Prudhoe Bay, Alaska. During the Cretaceous, Alaska would have had a temperate climate but dinosaurs living there would have to have dealt with up to six months of darkness. He believed that the distance necessary for migration would be too great and concluded the dinosaurs were permanent residents in Alaska. Therefore, those high latitude dinosaurs should have had a big survival advantage over other terrestrial animals in the dust-induced "lights out" scenario proposed by Alvarez, et al.

Polar dinosaurs are known to have existed during the Mesozoic and would have had to deal with a sunless environment for three to six months of the year. This assumes, of course, that the obliquity of the Earth was what it is today. Knowing that the structure of their brain indicated an adaptation for low light conditions (i.e., very large optic lobes), it can probably be assumed that these polar dinosaurs did live in periods of extended darkness and did not migrate to warmer climes in the winter. Again, a nuclear winter scenario should have favored the polar dinosaurs over those K-T survivors often pointed out, namely snakes, crocodiles, birds, amphibians, lizards, turtles, frogs, champsosaurs, etc., but it did not. Why not?

Dr. J. Keith Rigby, Jr., paleontologist at the University of Notre Dame, found dinosaur fossils above the iridium clay layer at Fort Peck, Montana. The presence of pollen from the Paleocene flora convinced him that dinosaurs had passed through the K-T boundary. Researchers also found a fossilized hadrosaur femur 1.3 meters above the K-T boundary in the Ojo Alamo Sandstone at the San Juan River of the Hell Creek Formation. This would indicate dinosaurs existed for at least 40,000 years above the K-T boundary. Critics claim that

24

the fossils found were "reworked" (meaning they were redeposited to a higher level).

As mentioned earlier, the Impact Theory of Extinction cannot explain why the above mentioned survivors of the K-T extinctions were able to survive the nuclear winter scenario. An acid rain environment is also suggested by that theory and those previously mentioned known survivors should have been negatively affected by the pH changes that acid rain would entail.

In 2004, the fossil remains of small dinosaurs (named Oryctodromeus) that burrowed and nested in the ground were found by scientists in southwest Montana, USA. The 95 mya bones were from an adult and 2 juveniles. They were unearthed in a chamber 2.1 meters long. Since the chance of finding the remains of such a dinosaur in its underground abode is extremely small, the chances are that there were many others that lived this way. If the impact theory is correct, these burrowing dinosaurs would have had a very high probability of surviving the K-T transition; yet none did. Why not?

Today, we know how difficult it is to grow tropical plants in the northern hemisphere. There are planting zones that show us the limits of where plants can thrive based on the temperature gradient spanning those zones. Just a few hours of a severe drop in temperature can kill a plant. Yet, during the K-T transition, tropical plants survived very well compared to northern plants. If the nuclear winter scenario is correct, how could that be?

Recently, study of the forams (foraminifera) near the K-T boundary by Gerta Keller, a Princeton University paleontologist, has indicated that the impact occurred about 300,000 years before the K-T boundary. The K-T boundary is defined by many paleontologists, not by the impact as some might think, but by the transition from Cretaceous foraminifera to the Cenozoic (or Tertiary) foraminifera. The new earlier date for the impact has called into question the alleged extinction effects of the impact.

PART III: THE VOLCANIC THEORY OF EXTINCTION

CHAPTER 5: THE OPPOSITION GATHERS

The 1980 publication in the journal *Science* of the asteroid impact theory of extinction was not met with unanimous approval by paleontologists and others in related fields. Although there were those who resented the suggestion that an extraterrestrial visitor could have had a major influence on Earth's history, some were more incensed by the intrusion of experts from another scientific discipline, especially by the physicist Luis Alvarez.

There was legitimate concern about whether the Alvarezes had rushed to judgement by connecting the apparent iridium spike at the K-T boundary with a "sudden" mass extinction. The 1980 paper failed to specify the time frame that it considered the extinctions to have taken place. The iridium abundance chart in that document displays a time interval for the clay layer as "~5000y?" and "1-10y?". That was not the kind of precision one would expect from a theory that makes the bold connection between an impact and major extinctions. That left a lot of leeway for the anti-impactors to justify their theory.

The consensus of most paleontologists in 1980, based on the fossil record, was that there was a gradual decline in both terrestrial and marine species during the late Cretaceous, interwoven with a series of extinction pulses. This process appeared to continue right up to the K-T boundary and was a multi-million year process. With the 1980 publication in *Science*, the Alvarezes had thrown down the gauntlet. The anti-impactors now had to explain the final extinctions of the Cretaceous with a more realistic and hopefully, a singular cause.

THE VOLCANIC THEORY OF EXTINCTION

The temporal relationship between the eruption of the Deccan Traps in India and the K-T transition along with the long held belief that the extinctions were gradual was the deciding factor for the anti-impactors to embrace volcanism as the extinction mechanism. The case for volcanism as the primary cause of the extinctions was, and still is, a powerful one.

The Deccan Traps volcanism spanned the K-T boundary. Estimates vary widely as to the total duration of the eruptions ranging from 300,000 years to 2,000,000 years. It is accepted that the eruptions were intermittent based on at least 100 distinct lava flows.

DECCAN TRAPS, India

The Deccan eruptions in India were the result of what is called a "hot- spot" volcano. Unlike the more commonly known volcanoes, such as Mount Saint Helens, which are formed from tectonically driven subduction processes, the hot-spot volcanoes are formed from lava flows, or plumes, that originate deep within the Earth near the core/mantle boundary. This gives them the potential for exuding much greater quantities of lava. The chemical composition of their lava is different from that of the non-hot spot volcanoes. It is basaltic in composition, hence the name "flood basalts" is applied to their outpourings. It is believed that the plumes, unlike the tectonic plates that float on the Earth's asthenosphere, are in fixed positions within the Earth's mantle.

At the time of the K-T transition, India was on a tectonic plate which was located near the midpoint, and off the east coast, of Africa. It was probably on the fastest moving tectonic plate and was moving toward its eventual rendezvous with the Eurasian Plate. The Himalayas were formed by that collision. The plate that India was on had already started to pass over the hot-spot that is known as Piton de la Fournaise located beneath Reunion Island and the Deccan eruptions were well underway prior to the K-T transition. The resultant lava flows would eventually blanket western India with 2 million cubic kilometers of lava.

The pro-volcanism theorists had a lot of ammunition with which to combat the pro-impactors. As mentioned, paleontologists had recognized a gradual extinction leading up to the K-T boundary for both terrestrial and marine fauna. This followed the widely accept *Uniformitarianism* principle which explained Earthly changes as long-term repetitive or cyclical Earth-bound processes.

After the 1980 *Science* publication, much criticism followed. The dark, one centimeter clay layer at Gubbio that had the one to two meter gray/white limestone below it stood in sharp contrast to the rest of the pinkish limestone. This anomaly indicates that something unusual was happening long before the clay layer was deposited. The paper didn't offer any explanation. And, as mentioned previously, the iridium graph was in conflict with the results of subsequent testing.

The clay layer at Gubbio was one centimeter thick but at other locations it was much thicker. Even at Stevns Klint in Denmark, one of the other primary sites analyzed for the 1980 report, the clay layer known as the Fish Clay was up to 45 cm thick with an average thickness of 4 to 6 cm. How could an instantaneous event such as an asteroid impact be responsible for some of those thick clay layers? It seemed more logical to believe that whatever produced those thick clay layers must have been acting over a very long time span.

The pro-impactors claimed that only an extraterrestrial impact could deliver the iridium found at the boundary. The pro-volcanism group countered that iridium has been emitted, and is currently emitted, by the Hawaiian Island volcanoes and other hot-spot volcanoes including the one that formed the Deccan Traps at Reunion Island. The implication, made by the pro-volcanism supporters, is that the most intense phase of the Deccan volcanism would have to have been concentrated at the much narrower time period of the K-T boundary window in order to obtain the iridium peak observed. And, the later iridium analysis by Rocchia and Crocket that indicates rising iridium levels before the peak has to be explained.

The pro-impactors strengthened their case by stating that the planar deformations of shocked quartz found at various K-T sites worldwide could only have come from an impact. They claimed the type of deformation pattern could be found only at other known impact sites and nuclear bomb test sites. This seems to be one of the biggest obstacles for the pro-volcanists. Their only explanation would have to be that the intense phase of the Deccan volcanism had to include pyroclastic or explosive episodes far more powerful than that of non hot-spot volcanoes. Since no hot-spot volcanism anywhere near the power of the Deccan eruptions has occurred in the last 65 million years, it is impossible to verify the shocked quartz deformation aspect of the pro-volcanism theorists.

CHAPTER 6: THE ENSUING DEBATE

In August 1983, an article appearing in the New York Times entitled *'Dinosaurs: Catastrophic   Theory Is Contested At Hell Hollow'*, written by John Noble Wilford, described the field work done by Dr. William A. Clemens, a professor of paleontology at the University of California at Berkeley. His work was at Hell Hollow, an area below Fort Peck in the Badlands of eastern Montana. Not finding dinosaur bones within 10 feet of the clay layer, Clemens remarked:

"You've got to have more than circumstantial evidence to prove the asteroid hypothesis."

The area in question has numerous fossil remains of Triceratops and was also near the great Western Interior Seaway in the United States that underwent a massive regression in the latter part of the Cretaceous. It is also one of the few areas where one can view a relatively complete stratigraphic transition over the K-T. Clemens made the following comparison about the missing fossils:

"That's like the 18 ½ minutes of the Nixon tapes....it's blank, no dinosaurs, almost no fossils at all. It's a mystery."

A coworker of Clemens, Carol Hotton, did uncover something that seemed to support the impact theory of extinction. She found that there was a sharp break in the fossil pollen near the end of the Cretaceous Period. This could be just a local phenomenon. However, the article also points out that many bones of champsosaurs, reptiles that look a lot like small crocodiles, are found before and after the K-T transition. Dr. Howard Hutchinson, of the Berkeley Paleontology Museum, noted the diminishing size and diversity of turtles in the late Cretaceous Period.

THE VOLCANIC THEORY OF EXTINCTION

> The diminishing size of the turtles mentioned above and the extinction of the largest (protostegid) turtles would be consistent with the GTME.

In October 1985, an article appeared in the New York Times entitled '*Dinosaur Experts Resist Meteor Extinction Idea*', written by Malcolm W. Browne, about the controversy. The article describes the polarizing effects of the debate and how publication of opposing viewpoints had been stifled. Those in the scientific community who voiced opposition to the impact theory had their careers jeopardized and risked having their grant proposals rejected. There was a kind of "pro-impact glass ceiling" effect in play as far as their careers were concerned. The Cold War was still underway and anyone who voiced opposition to the "nuclear winter" scenario of the proposed impact theory was seen as a militarist who might not be concerned about the effects of nuclear warfare.

The article describes a survey of 118 paleontologists at the 1985 annual meeting of the Society of Vertebrate Paleontologists. Only 4 percent believed that a bolide (asteroid or comet) impact could have caused the K-T extinctions. Another 27 percent believed that there wasn't a sudden mass extinction at the K-T boundary. About 80% did believe that there was an impact around the time in question but wasn't the cause of the extinctions.

One of the earliest opponents to the impact theory of extinction was Dewey M. McLean, a geologist at the Virginia Polytechnic University. In the early 1970s he proposed that the K-T extinctions were the result of a volcanic greenhouse effect. He believed that the Deccan Traps volcanism had disturbed the Carbon Cycle causing a heat induced reproduction failure of the dinosaurs. It is known that for crocodiles, for example, the gender of their offspring is influenced by the temperature of their developing eggs. Extremes in

32

temperature skews the percentage of newborn to either male or female. McLean also attributed the decimation of marine life to the warming of the oceans and the pH change that would have been caused by the outgassing of sulfur dioxide from the Deccan eruptions. Opponents of this theory claim that the sulfur dioxide outgassing would have led to global cooling and there are no signs of major cooling at the time of the K-T transition. And, the outgassing would have been insufficient to cause a noticeable change in the oceans' pH.

Vincent Courtillot, director of the Institut de Physique du Globe de Paris, is also a proponent of the volcanic theory of extinction. He cites the Laki fissure eruption of 1783 in Iceland which caused a 5 degree centigrade drop in temperature the following winter in the northern hemisphere and caused famine and the death of 10,000 people. He states that the outpourings did not reach as high as the stratosphere and concludes that the Deccan eruptions, being much larger, would have had a much more devastating effect.

The bias of scientific publications, such as the journal *Science*, were alleged by many of the pro-volcanism supporters. In December of 1992, Dewey McLean wrote a letter to Dr. Daniel E. Koshland, Jr., the editor of the journal *Science*. In that letter McLean questions whether *Science* has used a balanced approach to publishing articles on both the pro-impact and pro-volcanism theories. He writes that since the original 1980 publication of the Alvarez Impact Theory, *Science* published 45 pro-impact papers and 4 non-impact papers. One *Science* staff writer is alleged to have written to a pro-volcanism writer:

> **"Don't bother sending anything to Science which argues for a gradual extinction scenario. If it does not support impact, it will not get past our editor's desk."**

PART IV: THE GRAVITY THEORY OF MASS EXTINCTION (GTME)

CHAPTER 7: OVERVIEW

The search for an explanation of the massive size of the Mesozoic dinosaurs was the driving force for the development of this theory. As the author of this theory, I had muddled over the size paradox for many years. It was only after my career as a computer systems developer ended that I turned my full attention to this question.

My first thoughts were probably like those who also pondered this question and concluded that gravity had to be less than what it is today. But how could that be? Like others, my initial thought was that the Earth's diameter had changed; had gotten larger perhaps with the buildup of extraterrestrial debris raining down over millions of years. The gradual accretion of mass on the Earth would, according to Newton's Universal Law of Gravity, gradually increase the weight of all surface objects. This seemed like the only possible explanation although the increase in mass didn't seem like it could be substantial enough to explain it. After reading a book on geology and learning about the formation and breakup of the super-continent Pangea, a light bulb went on. Since the dinosaurs came into being after the Pangean super-continent was formed and went extinct as it broke apart, that nexus had to be the key to solving the problem!

I then set out to explain how the formation of Pangea could lower the surface gravity on the Earth. The result of that thought process resulted in the initial theory which was published (on the Internet) in 2004 and is reproduced in *Chapter 8: The Birth of the Theory*.

THE GRAVITY THEORY OF MASS EXTINCTION

I had confidence that my theory was correct. However, I knew that I had to go further in order to provide circumstantial evidence needed to sway anyone who might not take the theory seriously. And, not wanting to be labeled as just another "leaping unknown" with another crackpot theory, I continued my research. What followed in about a year and a half was the writing of Part 2 of the theory which was published in 2006 (also on the Internet) and is reproduced in *Chapter 9: The Evolution of the Theory.* That chapter describes the various Mesozoic life forms that had undergone gigantism and extinction, primarily the sauropods, mosasaurs, plesiosaurs, ammonites and forams.

After publishing Part 2, I started to think about whether the explanation I had given concerning the cause for the lowered gravitational effect could account for the apparent magnitude of the gravitational change. After thinking about this for some time, another light bulb went on. I realized that if the iron core(s) at the Earth's center had shifted away from a central position, and away from the location of Pangea, it could cause a profound decrease in the surface gravity on the super-continent. That core shift would have to have been a response to the coalescing of the continents on one side of the globe. The reduced gravity would be due to the fact that Newton's gravity law specifies an inverse relationship between the surface gravitational force (i.e., weight) and the square of the distance to the center of mass of the Earth, which would be near the center of the shifted iron core(s). This would have a much greater effect on surface gravity than an expanding Earth scenario.

In order to test the validity of this new upgrade to the original theory, I had to find something that would be uniquely explained by the core shift concept. Knowing that the core shift would create a gravitational gradient where the lowest gravity would be on the Pangean surface furthest from the shifted core(s) and therefore near the central, or equatorial region, I concluded that the largest dinosaurs, the sauropods, would probably have been concentrated in this region. This turned out to be fairly accurate. The result was

Part 3 of the theory, published in 2007 (also on the Internet) and is reproduced in *Chapter 10: The Eureka Moment of the Theory.*

Part 4 of the Gravity Theory of Mass Extinction theorizes that the Permian-Triassic extinction (~251 mya) was initiated by the low surface gravity that occurred when Pangea had formed. This would have released methane from the methyl hydrates at the bottom of the sea. Part 4 was published in 2007 (also on the Internet) and is reproduced in *Chapter 11: The P-T Extinction- A New Explanation.*

The end-Triassic (a/k/a Triassic-Jurassic) extinctions, also attributable to gravitational change, are described in *Chapter 13* of this book.

Many aspects of extinctions, K-T and others, which seem to be unexplainable based on currently accepted theories are readily explainable based on the gravitational change caused by the shift in the Earth's core. They will be addressed throughout this book.

Note that the core shift described in this theory includes both the solid inner and outer liquid iron cores. While it is easy to visualize the movement of the inner core, the liquid outer core might be more difficult. However, the lower mantle is of lower viscosity than the outer mantle allowing movement of the entire core. It is also assumed that the inner core moved over a greater distance than the outer core (i.e., the inner core did not remain at the center of the outer core). Currently, the inner core revolves at a rate slightly faster than the Earth itself. It is possible that this rate was reduced during the time periods analyzed in this book.

THE GRAVITY THEORY OF MASS EXTINCTION

THE BASICS—THE GRAVITY THEORY OF MASS EXTINCTION (GTME)

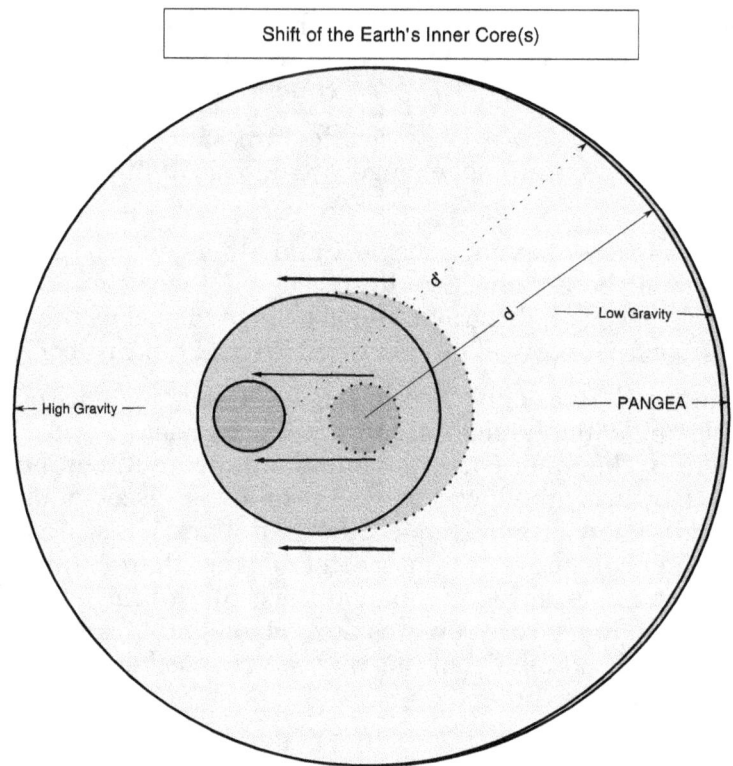

FIG. 7-1 Shift in the Earth's inner core(s)

The GTME can be visualized with reference to Figure 7-1. As the continents coalesced during the late Carboniferous and Permian Periods, the Earth's inner/outer iron cores shifted away from a central position and away from Pangea. This core shift increased the distance (from d to d') between the Earth's shifted center of mass and the Pangean surface, and therefore based on Newton's Law of Gravitation, lowered the force of gravity at the surface:

$$F = KMm/d^2$$

F is the force of gravity (i.e., weight) of mass m at surface of Pangea

K is a constant (sometimes named G)

m is the mass of an object at surface

M is mass of Earth
d is distance between the center of masses of M and m

The core shift is a response to the unbalancing effect of the consolidation of the continents on one side of the globe. The lowered gravitational field led to the gigantism of not only the dinosaurs but marine animals and flora as well.

During the Mesozoic Era, a period of some 185 million years, Pangea broke up gradually. Each movement of a continental tectonic plate had an effect on the movement of the inner/outer cores and most of those movements acted to return them toward an Earth-centric position. Most rapid major continental plate movements caused a pulse of higher gravitation at the surface of Pangea. These gravitational pulses caused corresponding extinction pulses.

THE GRAVITY THEORY OF MASS EXTINCTION

The core shift would, according to the GTME, explain why the most massive flood basalt volcanic episodes occurred during the Mesozoic Era. Flood Basalt Volcanism (FBV) is believed to originate at the Earth's outer core/mantle boundary. The shifting iron core would exert higher pressure on the part of the liquid iron core in the region ahead of its movement when it was moving away from its central location producing the Siberian Traps (250 mya). When it was returning toward its central location during the lengthy breakup of Pangea, the core movements produced the Deccan Traps (66 mya) and the North Atlantic Igneous Province (57 mya). The outpouring of massive FBV has tapered off considerably since then and there is no known FBV prior to the Siberian Traps. This is the basis for the author's belief that the Deccan volcanism was not a major cause of the K-T extinctions but a side effect of the gravitational basis for the extinctions.

Based on the linkage of the core shift and tectonic plate movement, as described by this theory, the following is possible. The most pronounced breakup and dispersal of Pangea at the end of the Mesozoic Era was a much faster process than its formation during the Permian, on a relative geologic timescale. There is the possibility that a more pronounced wobble of the Earth might have taken place at the end of the Mesozoic because of this rapid movement of the Earth's core(s). If that wobble happened, the orbits of asteroids that might have been near the Earth could have been disturbed. This might explain the coincidence of the Chicxulub impact at the K-T boundary. That wobble might have caused other, currently undiscovered, bolide impacts at that time.

The next three chapters contain the original theory as it was published on the Internet from 2004 through 2007. It must be noted that the original focus of the theory was on the gigantism of the Mesozoic dinosaurs, which led to evolution of the broader mass extinction theory described in this book.

CHAPTER 8: THE BIRTH OF THE THEORY

What follows is the part 1 of 4 of the original theory, published in 2004 on the Internet.

This document describes a new theory concerning the ascendancy and the extinction of the dinosaurs during the Mesozoic Era of approximately 225 to 65 million years ago. The geologic time scale of the Mesozoic Era is subdivided into three periods. These periods are:

TRIASSIC PERIOD about 225 to 193 million years ago. During this period the earliest dinosaurs appeared, displacing many of the existing reptiles.

JURASSIC PERIOD about 193 to 136 million years ago. During this period the diversity and physical size of dinosaurs increased tremendously. The larger terrestrial dinosaurs such as Allosaurus, Brachiosaurus, Diplodocus and Stegosaurus appeared.

CRETACEOUS PERIOD about 136 to 65 million years ago. During this period, especially in its early segment, the large dinosaurs were gradually replaced by smaller dinosaurs such as the boneheaded Pachycephalosaurus, Triceratops and Oviraptosaurs. By the end of this period, all of the known non-avian dinosaurs appear to have become extinct.

Many theories have been introduced to explain the extinction of the dinosaurs. Briefly, some of these are:

- A large extraterrestrial body (such as a meteor) struck the Earth's surface around 65 million years ago at the

THE GRAVITY THEORY OF MASS EXTINCTION

K-T (Cretaceous/Tertiary) boundary throwing up a worldwide cloud of soot and debris. This effectively caused a dramatic cooling of the Earth's surface which created an environment in which the dinosaurs could not survive.

• Massive worldwide volcanic activity occurred at the K-T boundary due to the Earth's increased plate tectonic activity, having the same effect as an extraterrestrial collision.

• The widespread expansion of flowering plants and trees caused the blanketing of the Earth's atmosphere with pollen creating an environment that was hostile to the dinosaurs.

• Massive cosmic ray activity destroyed the dinosaurs.

A NEW THEORY

The theory put forth in this document is a new one, vastly different from those that have been presented to date. This theory is:

The Earth's continents merged into a single massive continent prior to the Mesozoic Era. This super-continent is known as Pangea (or Pangaea). This continental consolidation and attendant mountain building, resulting from plate tectonic collisions, created an environment where the ambient gravitational force at the lower elevations and flood-planes of the super-continent was much lower than it is today. This permitted the unprecedented physical growth of the dinosaurs. As the continents broke away and drifted apart during the Mesozoic Era, the ambient gravitational field gradually increased, basically reversing the trend in physical size of the dinosaurs. This downsizing process, along with other factors, allowed mammals to increase in size and numbers, eventually displacing the non-avian dinosaurs.

THE GRAVITY THEORY OF MASS EXTINCTION

Isaac Newton postulated the Law of Gravity approximately 340 years ago. He theorized that any two objects exert an attractive force on each other. The magnitude of this attractive force is directly proportional to the product of the masses of the two objects and is inversely proportional to the square of the distance between them. Thus, for objects with masses M and m that are a distance d apart, the force F of attraction between them, according to Newton is:

$$F = KMm/d^2$$ **K** is a constant (usually called G)

According to this relationship, if the mass of one of the objects, M or m, were doubled, then the force of attraction F would be doubled. However, if the distance d between the objects were doubled, the resultant force F would be ¼ of its prior value. If the Earth is used as one of the masses (say M) and an object on the surface of the Earth is the second object m, then the force F can be considered the weight W of the object m.

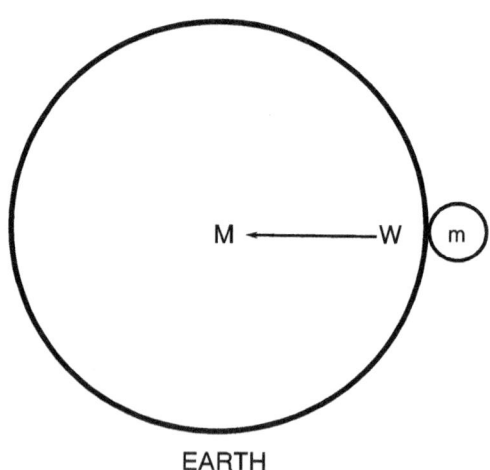

EARTH

The application of Newton's Law of Gravity in the above example is a little more complicated than it might seem for the following reasons:

- Mass m on the Earth's surface is attracted to every particle within the Earth. Therefore, the attractive force W between object m and the Earth (i.e., m's weight) is a vector sum of all of the m/particle forces.

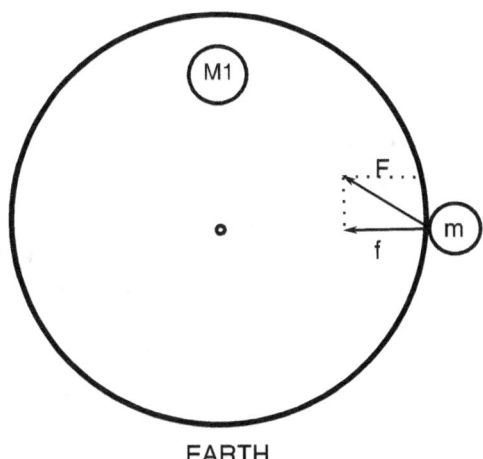

EARTH

THE GRAVITY THEORY OF MASS EXTINCTION

In the above example, object m and M1 (a random particle within the Earth's sphere) have a mutual attraction F. However, only the vector component of F (i.e., f) directed toward the Earth's center contributes to the weight of object m. The summation of all these vector forces between mM1, mM2, ... will compute the accurate weight of m. This is a calculus problem for a mathematician.

- Another condition which complicates the calculation of m's weight is the fact that the earth is not a homogeneous sphere. The Earth's interior consists of the crust, asthenosphere, transition zone, lower mantle, outer and inner core, all consisting of materials of different densities.

- Also, the Earth is not a perfect sphere. It is flattened at the poles and therefore, the force of gravity is greater (i.e., an object's weight is greater) at the poles.

If object m could descend into the Earth toward the Earth's center, its weight would decrease until it became weightless at the Earth's center. As the object descended from the Earth's surface, the Earth's mass above the object would exert an attractive force tending to pull the object toward the surface. The more powerful Earth centered force would offset the surface directed force until the object reached the Earth's center.

By the time the dinosaurs came into existence, the Earth's continents, as we know them today, were combined into a super-continent called Pangea. The process of continental drift, which is still occurring, caused the continental land masses to combine. During this process of land mass collision, huge mountain ranges were uplifted. The lowland areas and tidal basins were dwarfed by these towering mountain ranges.

The effect of the consolidation of the Earth's land masses and the creation of tall mountain ranges and volcanic peaks lowered the strength of the gravitational field in Pangea, especially at the lower elevations. This is comparable to an object being lowered into a hole toward the Earth's center as described in the previous section "Review of The Earth's Gravitational Field." As mentioned, the Earth's mass above the object would partially cancel the attractive force of the Earth within the sphere below the object.

The lowered gravitational field permitted the flora and fauna of Pangea to grow to a size which would not be possible today. Conifers that grew over 100 feet tall and at least 5 feet wide were common. The remnants of these trees can be found in the Petrified remains in the Yellowstone National Park. The remains of the giant dinosaurs of the Mesozoic Era can be found in museums throughout the world.

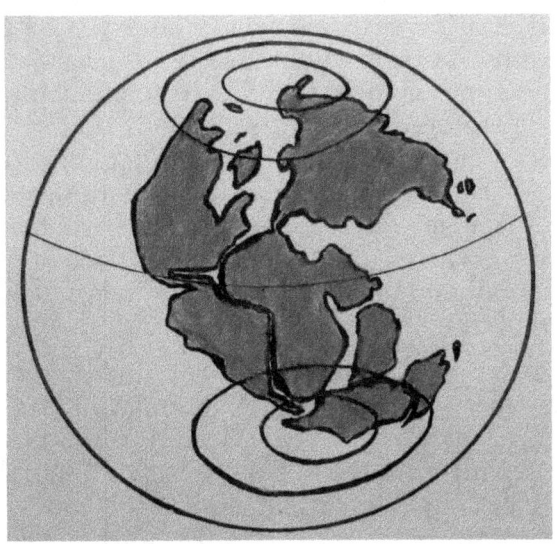

Super-continent Pangea At The End Of The Triassic Period

THE GRAVITY THEORY OF MASS EXTINCTION

If one studies a cladogram, or family tree, of the dinosaurs, one can see the evolution of dinosaurs during the Mesozoic Era. During the Triassic Period, dinosaurs started out as physically small specimens such as Coelophysis, Herrerasaurus and Eoraptor. Larger and larger dinosaurs evolved during the Jurassic Period including the sauropods, stegosaurs, allosaurs, etc. They were dominant in the regions where the gravitational force was the lowest. Many of the largest species of dinosaurs, including the few named above, became extinct during the early Cretaceous Period (from about 136 to 100 million years ago.) The reason for this will be explained next.

The super-continent Pangea began to break apart during the Triassic Period. As the continents separated, the ambient gravitational field increased. It would be expected that the large sauropods would be most vulnerable to an increase in gravity. It would become impossible for their hearts to pump blood up to their lofty skulls. Therefore, the giant sauropods such as Apatosaurus, Brachiosaurus, Camarasaurus, Diplodocus and Barosaurus became extinct in the early Cretaceous period. Since the continents which separated from Pangea retreated at different rates, the demise of the large dinosaurs occurred at different rates in each of the continents. Smaller sauropods and hollow boned theropods, would survive for a longer period of time in the Cretaceous.

The pterosaurs, the flying reptiles, underwent evolutionary change to compensate for the gravitational increase. The longer forearmed (i.e., longer winged) pterodactyl (Pterodactylus) supplanted the shorter forearmed pterodactyl (Rhamphorhynchus). But it eventually became extinct because it could not continue adapting. The premise on which the movie *"Jurassic Park"* was made is a bit faulty. If we were able to hatch eggs of the large dinosaurs today, they would have a very short life span. The pterodactyls would not be able to fly today.

The continental separation continued past the Cretaceous Period and the effect on existing species continued. Since there is always an interdependence of animal life, many different smaller and medium size dinosaurs gradually disappeared. The largest sauropods, with their enormous appetite, acted as road clearing machines in the Cretaceous forests. Other life forms, mammalian and dinosaurian, that flourished in those open areas would have been affected by the extinction of the largest dinosaurs. The larger theropod predators would also have been negatively affected by the disappearance of the large sauropods. This new environment permitted the growth, in size and numbers, of the mammals. The new mammals, with larger brains and ever increasing physical size became a threat to the egg laying dinosaurs. Only the avian dinosaurs, the forebears of today's birds were able to survive. Being able to lay eggs in remote places, inaccessible to the mammalian predators, was their key to survival.

There is no doubt that other environmental factors influenced the extinction of many of the dinosaurs throughout the Mesozoic Era. Changes in climate, impacts of extraterrestrial objects, changes in flora, etc., definitely affected the evolution and extinction of life on Earth. However the changes in the ambient gravitational field on the Earth's surface was the dominant factor in the extinction of the dinosaurs.

CHAPTER 9: THE EVOLUTION OF THE THEORY

What follows is Part 2 of 4 of original theory. It was published in 2006 on the Internet.

I. INTRODUCTION

This document is a follow-up to the original theory of Dinosaur Gigantism and Extinction entitled "THE RISE AND ALL OF THE DINOSAURS", copyright 2004. A brief summary of the original theory follows.

The existing continents of the world, through plate tectonic motion, coalesced into a single super-continent called Pangea approximately 260 mya. This coalescing process caused a gradual reduction in the gravitational field on the surface of the super-continent such that some species of land and aquatic animals (and flying reptiles) were able to grow to extreme physical size that would not be possible in today's gravitational field. The breakup of Pangea resulted in an increasing gravitational field which led to the downsizing of many land and sea animals and was the primary reason for the extinction of dinosaurs as well as flying and sea reptiles. A thought experiment, which might clarify how this happened, follows.

Imagine that you are standing somewhere on the Earth's surface. At the opposite side of the planet are two very tall buildings. Those two buildings start moving toward you but in an opposite direction from each other. In other words, one building will eventually reach you, for example, from the west and the other from the east. In their initial position, their effect on your weight will be insignificant because they are very far away; a distance equal to the Earth's diameter. The buildings move around the planet until they appear on the horizon. This is the point at which their effect on your weight starts to become significant for two reasons. Before we address those two reasons, let's visualize the final position of those two buildings.

THE GRAVITY THEORY OF MASS EXTINCTION

That position has you touching one building with your right hand and the other building with your left hand. It should be clear that the effect on your weight of those two buildings went from a positive one (they supplemented the Earth's pull on your body) to a negative one (they are above you cancelling part of the Earth's pull on your body). As the two buildings appear on the horizon and move toward you, the two factors that come into play are:

1. The distance from you to the building gets smaller. Newton's Universal Law of Gravitation (see original theory) posits that the gravitational force (i.e., weight, in this context) is inversely proportional to the square of the distance between two objects. Thus, the impact of those two buildings on your weight rapidly increases as they approach you.

2. The two buildings start to direct their gravitational effect on you away from the mid-point of the Earth (which increases your weight) and toward a point above you culminating in the maximum negative effect when they are next to, and above you.

When the buildings are replaced by the Earth's continents coalescing to form Pangea, it can be concluded that the gravitational change, and therefore the reduction in the weight of objects on the super-continent, would have been significant.

During the approximately 125 million years that the Laurasia/Gondwana super-continent existed, dinosaurs flourished. Note that Pangea is the name applied to the consolidation of these two large land masses, which in turn consist of a consolidation of smaller continental land masses. Many of the dinosaurs grew to astounding size. The breakup of Gondwana and the subsequent splitting of Africa and South America approximately 135mya was followed by the disappearance of the largest of the sauropod dinosaurs during the early Cretaceous Period. Referencing the previous thought experiment, it should be apparent that the greatest rate of change in gravitational force would occur at the inception of the breakup of the largest land masses of Pangea. It has to be noted that the breakup of Pangea was not a simple separation of all the present-day continents occurring at the same time. The time line of Pangea, scientists believe, is approximately:

420-395mya The continents merged into 2 super-continents: Laurasia...consisting of Antarctica, Australia, N. America and India and Gondwana ...consisting of Africa and S. America.

350-260mya Laurasia and Gondwana combine to form Pangea.

248mya Pangea starts to breakup into Laurasia and Gondwana.

180-160mya Gondwana broke apart separating into the Africa/S. America land mass and the Antarctica/Australia/India land mass.

145-135mya S. America splits into a continent separate from Africa. N. America splits from Laurasia.

100-96mya Australia and New Zealand separate from Antarctica. Also, Madagascar and the Seychelles Islands separate from India.

76mya New Zealand separates from Australia.

THE GRAVITY THEORY OF MASS EXTINCTION

This timeline helps to explain why the largest sauropod dinosaurs and the largest reptiles became extinct by the mid-Cretaceous (approximately 110-100 mya). By this time, all large present-day continents had separated and were drifting apart. This was the time of the greatest rate of increase of the gravitational field on the continents. Could Reduced Gravity Growth (RGG) land animals evolve to a smaller size to compensate for an increasing gravitational field? In most instances they could not. Their predators had simultaneously evolved to a RGG size sufficient to successfully prey on those species. There could only be a negative impact resulting from downsizing.........increasing predation and eventual extinction. The increasing gravitational force would have a profound effect upon all land, sea and flying animals. It was this destabilizing change in force (i.e., weight) and the resultant extinctions that created gaps in the diversity of dinosaur species on land that permitted carnivorous mammals to rise from small shrew-like rodents to larger animals capable of competing with, and eventually displacing all, non-avian dinosaurs.

II. THE K-T EXTINCTION----------- VOLCANISM, IMPACT OR GRAVITATIONAL?

Paleontologist David Raup gave his criteria for mass extinction:

"For geographically widespread species, extinction is likely if the killing stress is one so rare as to be beyond the experience of the species and thus outside the reach of natural selection."

Only this Gravitational Theory conforms with this criteria. Volcanism and bolide impaction have been common occurrences during the 160 million years that the non-avian dinosaurs inhabited the Earth.

Many scientists have focused exclusively on volcanism and impact phenomena to explain the K-T Extinction and prior major extinctions. I believe future research will prove that none of the major extinctions were the result of these two activities. The coalescing and breakup of continental land masses will be found to be the culprit. Specifically, the concomitant alterations in the gravitational field caused the direct extinction of certain species that had developed RGG structures that were incompatible with a significant change in gravitational forces. Extreme volcanism is a concomitant action of continental collisions and breakup so that finding volcanism activity at a time of major extinctions is best described by the Latin phrase:

"Cum hoc ergo propter hoc"

The above describes the mistaken supposition that when two events occur at the same time, one must have caused the other. Major impacts have not been found to correspond with major extinctions except for the K-T extinction. However, it must be noted that some scientists question whether the Chicxulub crater is a result of bolide impaction. This will be addressed later in this document.

I also believe that attempts to establish a precise periodicity to major extinctions cannot be established. The reason for this is that the coalescing and rifting of large land masses due to plate tectonics, which in my opinion are the primary cause of the extinctions, do not occur in intervals of equal duration nor is the pattern of consolidation and breakup the same each time.

III. WHY DID SOME SPECIES SURVIVE THE K-T EXTINCTION AND OTHERS DID NOT?

The phrase "K-T extinction event" will not be used here because there was no single event that caused the demise of all of the non-avian dinosaurs and other aquatic biota. The K-T boundary is an arbitrary period in time which appears to delineate the transition from dinosaurs to mammals. Because this transition was a gradual one, it will not be surprising to find some residual dinosaur fossil remains above this boundary. This could happen but the absence of dinosaur remains in the so called "ghastly three meter gap", the three meter expanse below the clay boundary layer, would make it a slim possibility.

-LAND SPECIES

When the Pangean super-continent was formed, the reduced gravity environment raised the physical-limit ceiling for both flora and fauna. Survival-oriented evolution would dictate the growth patterns of the Earth's biota. Some land animals would hardly be affected in terms of their physical size. Others would take the Reduced Gravity Growth (RGG) path to gigantism, evolving to sizes not possible today. Some of the largest dinosaurs, all sauropods were:

THE GRAVITY THEORY OF MASS EXTINCTION

Name	weight	length	time period
1. Argentinosaurus	100 tons	120 ft	100-93mya
2. Paralititan	75 tons	100 ft	94mya
3. Brachiosaurus (40-50ft tall)	30-80 tons	85 ft	156-140mya
4. Sauroposeidon (60 ft tall)	55-65 tons	100 ft	110mya
5. Supersaurus (54 ft tall)	55 tons	138 ft	155-145mya

Note that the weight given above are the estimates of the dinosaur's weight in today's gravitational environment. If those sauropods could have been weighed when they were alive in the Mesozoic Era, one would have to divide the above weight by a factor of 4 to 6. Not all land animals would be affected by RGG. It is for this reason that they passed through the K-T boundary relatively unaffected. Some of those were:

- Most mammals
- Snakes
- Turtles
- Many small dinosaurs (i.e., birds)
- Crocodilians
- Lizards

It has to be noted that within the Crocodilian group, there was a sub-species that did take the RGG path to gigantism. This was the case with the "Super Croc" whose remains were found in the Sahara Desert. Named Sarcosuchus imperator, the ancient reptile lived approximately 100mya and weighed as much as 10 tons and was about 40 feet long. Super Croc went extinct while its non-RGG cousins are with us today. It is no coincidence that the largest dinosaurs became extinct in the early to middle Cretaceous along with the largest terrestrial reptiles such as Sarcosuchus imperator. The rifting of Laurasia/Gondwana into the present-day continents had been complete (approximately 135mya to 100mya) and the rate of change of the gravitational force was at its maximum.

If we think about the thought experiment described earlier, the rate of change of the gravitational field will become apparent. If an object is located directly below those two tall buildings, the buildings are having their maximum negative effect on the object's weight. As those two buildings move away toward the horizon, the component of their respective gravitational pull on the object away from the center of the Earth becomes smaller very rapidly because the angle between the radial line from the Earth's center through the object and the line from the object to one of the buildings is going from zero to ninety degrees rapidly. Thus a smaller component of the buildings' pull affects the object. In addition, the inverse square rule for gravitational force (see original theory) compounds the effect. Therefore, when Pangea began to split apart, the greatest rate of gravitational change was at the start of the breakup of the largest land masses. As mentioned earlier, the breakup of the super-continent happened in stages starting with Laurasia and Gondwana. This might explain the smaller extinctions, sometimes referred to as "stepwise extinctions", believed to have occurred during the Mesozoic. The RGG land animals would, if they could not evolve to a state compatible with an increasing gravitational environment, become extinct. Evolutionary downsizing is much more difficult to do than upsizing and this will be addressed later in this document.

The non-RGG land animals and those that were only slightly affected directly by an increasing gravitational environment would be affected indirectly. We know than even in today's world that the removal of a single species or sub-species can have a cascading effect. The gradual extinction of the RGG land animals during the Cretaceous had a disruptive effect on the dinosaur diversity equilibrium that had existed for over 100 my. It was this equilibrium and diversity that held mammals in check, relegating them to small, burrowing shrew-like animals. Small and medium size carnivorous dinosaurs were the obstacle to mammal growth and expansion.

In the late Jurassic, a period of sauropod gigantism, there were at least 6 long-necked brontosaurs that could feed tripodally (by rising up on 2 rear legs). Their heads were uplifted to 40 feet or more above the ground to reach the treetops. By the early Cretaceous all of these brontosaurs had become extinct and the sauropods that succeeded them were shorter. Their replacements had long necks but held them out more horizontally. How do paleontologists explain that? Today, we marvel at a giraffe, the tallest land-living mammal, that feeds at a mere 16-18 feet above the ground and does not feed tripodally. A giraffe's heart weighs up to 24 pounds and must produce twice the normal pressure for a large mammal in order to supply the brain with blood.

THE TITANOSAUR DILEMMA

As mentioned earlier, almost all of the largest dinosaurs, the sauropods, became extinct during the early Cretaceous Period. This extinction is consistent with the gravitational theory of gigantism and extinction described in this document. However, there is one lineage of sauropods that did continue to exist until the end of the Cretaceous. This would seem to be a major flaw of the theory. Why were these sauropods able to flourish in an increasing gravitational field while all the others had died out? I can only speculate as to the answer to that question. The fossil record for titanosaurs is not as expansive as it is for other dinosaurs. This might be because most of the titanosaur fossils found, and there haven't been a lot, have been found in South America and not in North America.

Here's why I believe the titanosaurs were able to thrive well into the Cretaceous Period:

1. In general, the titanosaurs were smaller than the earlier sauropods. With the exception of Argentinosaurus (with a length of up to 100 feet), the other titanosaurs didn't exceed about 60 feet in length. It should be noted that the remains of Argentinosaurus were found in the Cenomanian Age (about 96 mya) and not toward the very end of the Cretaceous.

2. The presence of armor, in the form of bony scutes on the neck and upper bodies, on many if not all of the titanosaurs is very important. Their presence could signify an attempt to downsize! As mentioned elsewhere in this document, downsizing for large sauropods was unlikely because their predators also followed the RGG path to gigantism. The development of armor could have been a way of compensating for the gravitationally induced pressure to downsize.

3. The "wide-gauge trackways" of the titanosaurs could imply several characteristics which would be in conformity with an increasing gravitational force:

- Titanosaurs were less migratory than other sauropods implying that they did not form herds. By evolving a lower, squatter form, their energy consumption needs would be much less.

- Titanosaurs were more likely to inhabit swamp-like areas so that their relatively large size would be less of a handicap due to the buoyancy of the lakes and rivers they spent most of their life in.

THE GRAVITY THEORY OF MASS EXTINCTION

- The recent discovery of a very massive titanosaur, Paralititan stromeri, in what was a mangrove swamp in Egypt does support the above speculation. Paralititan means "tidal giant." The estimate of 94mya, close to the time frame of Argentinosaurus, supports my belief that the titanosaurs became smaller, less mobile and acquired more armor as the end of the Cretaceous Period approached to compensate for the increasing gravitational field.

The almost complete extinction of the marsupials but not placental mammals at the end of the Cretaceous is directly explainable by the gravitational theory. The circumstantial evidence for a lowered gravitational force during the Mesozoic Era is overwhelming!

The disappearance of the RGG dinosaurs created gaps in the diversity of the dinosaurs. This is what confuses some who mention that mammals coexisted with dinosaurs for many tens of millions of years and therefore could not suddenly pose a threat to them. The belief was that mammals had remained at a small shrew-like size during the entire reign of the dinosaurs. Recent discoveries have falsified that belief. In China, a carnivorous, dog-size mammal from approximately 130 mya named Repenomamus giganticus was discovered. The fossil remains of a smaller, related species approximately 15 inches long, named Repenomamus robustus was also found. And, what is significant and very important is the fact that the remains of a juvenile dinosaur was found in the smaller mammal's stomach. This is undeniable proof that some mammals had become carnivorous adversaries of the dinosaurs capable of not only scavenging the eggs of dinosaurs but also preying on the their young long before the K-T boundary.

As a result of the increasing gravitational field, the gaps in the dinosaur diversity became wider. Fewer species of small carnivorous dinosaurs allowed for the rapid expansion of smaller carnivorous mammals. The dinosaurs, laying their eggs on the unprotected surface of the ground, would become vulnerable to the new, crafty, more intelligent, placenta-based mammals. Unlike the reptiles that survived the K-T extinction and exist today, such as crocodilians and turtles which bury their eggs below ground, the dinosaurs would gradually lose the survival battle with the mammals.

-SEA SPECIES

Many sea animals, as with land animals, would follow the path of RGG gigantism. The breakup of Pangea with the concomitant increase in the gravitational field strength would similarly affect sea animals which could not evolve to a form compatible with that change. It would be expected that the RGG sea animals that evolved as surface feeders would be affected the most. Their bodies evolved to give them three advantageous characteristics:

Speed

Buoyancy

Large body size

Plesiosaur

Plesiosaurs, for example, relied on these characteristics. There were different lineages of plesiosaurs but the one which was prevalent in the late Cretaceous Period was the Elasmosaurus. The largest ones were up to 46 feet in length with a neck almost 20 feet long. Its extremely long neck and small head allowed it to come within striking distance of smaller fish undetected. Clearly, its long neck was only possible with an oversized body. It could not possibly downsize and still be an effective hunter in the sea.

Ichthyosaur

Ichthyosaurs were a large dolphin-like swimming reptile built for speed (of about 40mph) that became extinct. But it became extinct about 90mya, long before the K-T boundary. Increasing gravity and not an event at the K-T boundary is the most likely cause. The Ichthyosaurs were around in the early Triassic and diversified throughout the Triassic and most of the Jurassic. Only one species, the Platypterygius lasted into the Cretaceous. More than likely, its method of feeding, that of high speed pursuit, became a disadvantage for ichthyosaurs as the gravitational field increased. It also gave live birth to its young, a method known as viviparous (or ovoviviparous). Usually this is done in surface waters and might also have been negatively affected by increases in gravitational forces.

Mosasaur

THE GRAVITY THEORY OF MASS EXTINCTION

Mosasaurs were large sea reptiles which apparently did not survive beyond the K-T boundary. They are believed to have existed from about 85-65mya. Described as the top predator in the sea, they seemed to evolve relatively soon after the Ichthyosaurs became extinct. They were up to 17 meters in length and up to 20 tons in weight. It is believed that they are related to snakes and monitor lizards. They were not fast swimmers and were ambush predators. This is probably why they survived into the late Cretaceous with the increasing gravitational field while the high speed pursuit Ichthyosaurs disappeared earlier. They hunted in near-surface waters.

One factor which might override all others relative to the sea reptiles is the following:

All sea reptiles must rise to the surface to obtain air to breathe. An increase in the gravitational force would have a negative effect on that function.

A second factor, mentioned above, is that they are viviparous, meaning that they give birth to their young alive. This is done in near surface waters. This is another function which would be negatively affected by an increasing of the gravitational field strength.

Smaller sea creatures were also vulnerable to the gravitational increase. In general, where a certain level of buoyancy was critical for survival and the sea life forms were not able to evolve to compensate for the increasing gravitational force, extinction was the result. This would be true of the molluscan ammonites, belemnites and planktonic life forms. Bottom feeders should have been the least affected by an increasing gravitational field. Gastropods, brachiopods, radiolaria, diatoms, dinoflagellates and bottom dwelling foraminifera were least affected.

Not every species, land or sea based, that disappeared toward the end of the Cretaceous can be attributed directly to the

66

gravitational change. The breakup of Pangea would affect species that thrived in the shallow waterways that disappeared as the super-continent gradually broke apart. Changes in water temperature and salinity could have influenced their extinction.

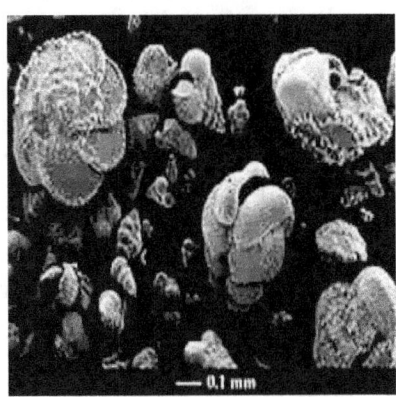 

| Foraminifera prior to K-T boundary | Foraminifera after K-T boundary |

Foraminifera prior to K-T
boundary

Foraminifera after K-T
boundary

Photo courtesy National Museum of Natural History

THE FORAMINIFERA CONTROVERSY

Foraminifera, or forams, are small single-celled protozoa that are widespread throughout the ocean. They float at various depths and many groups of the forams became extinct near the K-T boundary. Micropaleontologists have long debated whether most of the forams became extinct right at the K-T boundary or whether they started to gradually become extinct long before the boundary.

The gravitational theory of extinction that I have proposed supports the gradualist theory of foram extinction. As already pointed out, any life form that took the RGG gigantism path would become extinct when a major increase in the gravitational field

67

occurred. The species of forams that suffered a much lower rate of extinction at the K-T boundary was the benthic variety that lived on the sea floor. Some who support the Impact Theory of Extinction would say that they survived because they were the least exposed to toxic contamination of the upper surface waters of the ocean and were able to survive on the detritus on the bottom of the ocean. I disagree. As mentioned above, bottom feeders of any kind would be the least affected by an increasing gravitational environment. The buoyancy of forms in the upper layers of the oceans was compromised by an increasing gravitational field. Photos from the National Museum of Natural History show the size contrast between early Tertiary and Later Cretaceous (Maastrichtian) forams, the latter being much larger. Some paleontologists attribute this downsizing to the Lilliput Effect. I disagree with them. The decrease in size and the survival of bottom dwelling forams are, in my opinion, a direct consequence of the gravitational change described in the theory I have written.

Ammonite

THE AMMONITE CONTROVERSY

Ammonites are an extinct group of marine animals. Most had spiraled shells similar to the existing nautilus but are more closely related to the squid or octopus. Their shells resemble tightly-coiled rams' horns. A "squid in a snail shell" would be a good way to visualize them. Ammonites are believed to have lived in open seas and most, if not all, were not bottom-dwelling. It is believed that they were good swimmers with flattened, discuss-shaped shells.

They were able to control their buoyancy by being able, in effect, to pump water into their shell enabling them to ascend and descend in the water column. Few ammonites in the lower-mid Jurassic reached 9 inches in diameter. Larger forms were found in the upper Jurassic and lower Cretaceous. Some were two feet in diameter. The largest recorded in North America was four and a half feet in diameter.

The ammonites cannot be found above the K-T boundary and therefore the gradualists/catastrophists have debated the cause of the ammonite extinction extensively. Most experts agree that ammonite diversity was decreasing before the K-T boundary. Once again, I believe that the gravitational theory I have written readily explains this. As mentioned above, ammonites started out in the early Jurassic with shells about 9 inches in diameter and eventually grew to forms with shells over four feet in diameter. In other words, they followed the RGG path to gigantism as did many other sea and land creatures. They could control their buoyancy by pumping water in and out of their shells. When the gravitational field gradually increased, their RGG gigantism, vis-a-vis their large shells, became a disadvantage in terms of buoyancy and speed. Their cousins, the squid, did not have the weight of a shell to contend with. They are still around today. It should be noted that the nautilus did not become extinct during the period when the ammonites did. This might be because they had much harder shells than the ammonites and were able to rise and descend in the water column through a much wider depth range. In other words, they were better able to compensate for a changing gravitational field.

IV. ASTEROID IMPACT?

In writing the original Gravitational Theory of Dinosaur Gigantism and Extinction (*"The Rise And Fall Of The Dinosaurs"*), I did not address the Asteroid Impact Theory of extinction because I believe that an asteroid impact near the end of the Cretaceous, if it did occur, was purely coincidental. The proponents of this theory should be asked to explain the following questions:

1. If dinosaurs had flourished up until the impact, shouldn't we find the remains of some dinosaur herds, rookeries or even a single dinosaur in the southwestern USA covered with a fine layer of iridium enriched mud? There would have been sand storms, mud slides and tsunamis that would have buried numerous dinosaurs. Even though finding dinosaur remains are like looking for a needle in a haystack, why haven't paleontologists unearthed one fossil that meets this criteria?

2. Why would frogs, snakes, crocodiles, lizards and birds, all animals which would be highly sensitive to a drop in temperature as well as acid-rain pollution, survive the impact and its fallout?

3. If the impact caused a "nuclear winter", one would expect the tropical flora to be completely decimated. That doesn't seem to have happened. Why not?

4. If the asteroid struck at the Chicxulub location, and at a low angle as many believe, directed toward the southwest U.S., why haven't

70

numerous pieces of the asteroid been found? I know that normally a very large asteroid that impacts vertically is vaporized in an explosive manner, but it is believed that the K-T impact was not a vertical impaction.

Although the evidence for an impact at Chicxulub near the end of the Cretaceous seems very convincing, there are scientists that believe that the Chicxulub crater was formed by a lithospheric gas explosion caused by or related to the Deccan Traps volcanism in India. If they are correct, it would explain why the massive energy release of the theorized impact, which some estimate to be equivalent to the energy of 7 billion bombs the size of the one dropped on Hiroshima, doesn't seem to correlate with the damage expected. That gas explosion theory could explain the shocked quartz, iridium spike, tsunamis and related anomalies. It would also explain what I think is another peculiar coincidence. The alleged asteroid just happened to impact a point where tectonic plate activity was very active! Also, even though it is believed the crater is in an area that was a shallow sea at the end of the Cretaceous, seeing maps with the crater halfway under land and half under sea seems unusual.

V. A PREDICTION

The question arises as to how one could prove that a lower gravity environment existed during the age of the dinosaurs. It would seem obvious to a lot of people that land animals that rival the Blue Whale in size could not possibly live on the land surface of the Earth today. But how does one prove this in a scientific way?

Some have tried, using allometric studies of bone size and muscle tissue, to calculate what the maximum weight would be for a land animal today. Their estimate is about 20 tons, somewhat larger than that of the largest measurement of the present-day African elephant (13-14 tons). This is far smaller than the conservative weight estimate of 80 to 100 tons for the largest sauropods.

A good theory is supposed to be capable of supporting a prediction based on the fundamentals of that theory. The following is a prediction based on the theory that I have written.

The Deccan Traps are well-known volcanic remains in India. Plate tectonic studies tell us that when the Pangean super-continent broke apart, the Indian sub-continent drifted north from a position off the west coast of Africa where Madagascar currently is, across what is now the Indian Ocean on a collision course with the Asian continent. It drifted over a hot-spot, a deep volcanic plume in the ocean floor. The long lasting but intermittent volcanic eruption altered the surface of the Indian sub-continent with a flow of basaltic lava, thus forming the Deccan Traps seen in India today. Because this happened over a period of 6 to 10 million years and it also spanned the K-T boundary; we now have a natural gravity seismograph!

If geologists were to carefully examine the flow patterns of the lava along the entire length of the Deccan Traps, they should find a gradual pattern change going from a minimum at one end to a maximum at the other end. A change in gravitational force should affect the flow pattern of any fluid. I believe that verification of this flow pattern change would be sufficient to the scientific community to accept the gravitational change described in this theory.

VI. CONCLUSION

The Gravity Theory of Gigantism and Extinction that I have presented in the earlier (2004) document entitled *"The Rise And Fall Of The Dinosaurs"*, supplemented with the information provided in this document, I believe, is a viable explanation for the evolution and extinction of land, air and sea animals during the Mesozoic Era. I also believe that it may explain prior mass extinctions.

I believe that the debate between the gradualists/catastrophists, specifically the adherents of the volcanism/bolide dichotomy of extinctions have overlooked the most likely cause for the K-T extinctions, the gravitational one. Anomalies associated with either of the two popular theories of extinctions disappear with the gravitational theory of extinction. With the publication of this document, I believe the scientific community has enough information to begin to evaluate the theory I have presented.

CHAPTER 10: THE EUREKA MOMENT OF
THE THEORY

What follows is Part 3 of 4 of original theory. It was published in 2007 on the Internet.

I. INTRODUCTION

This document is the third in a series entitled "*The Rise And Fall Of The Dinosaurs.*" Parts I and II describe a new theory that explains the gigantism of dinosaurs during the Mesozoic Era and how it related to their eventual extinction. The theory posits that a gradual reduction in the Earth's gravitational field occurred as the terrestrial continents coalesced to form the super-continent of Pangea over 200 million years ago. The subsequent gradual breakup and drifting apart of those continental land masses was accompanied by an increasing gravitational field. This increase in gravity led to the extinction of many land and sea animals, especially the dinosaurs. The dinosaur/mammal equilibrium was disrupted, allowing the mammals to eventually displace the dinosaurs. This document supplements the prior two and introduces an additional factor to explain how the gravitational field strength could have been altered during the period in question.

II. OTHER GRAVITY THEORIES

Others have also come to the conclusion, based on the megafauna of the Mesozoic, that gravity must have been different during that time. Some have suggested that a large celestial body could have been in proximity to the Earth during that period and thereby acted to counter the gravitational field of the Earth. Some have suggested that the gravitational constant (the "G" in Newton's Universal Law of Gravitation) had changed. They suggest that it instantaneously decreased prior to the advent of the dinosaurs and then increased around 65mya.

The Expanding Earth Theory, initially proposed by Samuel Warren Carey in 1956,1976 was also used as a basis for a gravitational change. Carey hypothesized that because the continental outlines seemed to mesh, the Earth must have been much smaller, with an expansion of 33% since 200mya. Those who supported this theory, in its relation to reduced gravity, attempted to explain dinosaur gigantism on the basis that a smaller Earth would entail weaker gravity. Since Alfred Wegener's theory of continental drift has been almost universally accepted and explains the profile matching of the continents, the Expanding Earth Theory has been discarded by most scientists.

The theory described in this and the other two prior documents mentioned earlier offers a different explanation. And, that explanation is directly related to the formation and breakup of the super-continent of Pangea due to plate tectonic activity. Could the consolidation of the continents cause a substantial change in the surface gravity on Pangea? Since there doesn't seem to be another adequate explanation for the anomalous size of the Mesozoic fauna, one has to explain how the gravitational change could occur. One explanation was given in the prior two documents. Further study has added another factor to explain the gravitational change. The following section addresses that subject.

III. THE SHIFT OF THE EARTH'S CORE

The shift of the Earth's solid inner core or both the solid inner core and the liquid outer core must be considered. With the consolidation of the continental land masses in a relatively confined surface area of the Earth, a shifting of the core away from Pangea within the equatorial plane could account for a lowering of the surface gravity of Pangea. The shift of the core would act to maintain the center of rotation of the Earth along the same axis. Figure 10-1 is a representation of this situation.

A rough estimate of the change in the gravitational force can be made using Newton's Universal Gravity Law:

$W = GMm/r^2$

where:

W=weight of an object of mass "**m**" on Earth.
M=mass of Earth
r=distance from center of mass of Earth to
    location of "m" on Earth's surface
G= a constant

The resulting percentage decrease in gravitational force from a shift in core(s) is, as shown in Fig. 10-1:

(Current radius of Earth) squared

~ ───

(Distance from shifted center of mass to surface) squared

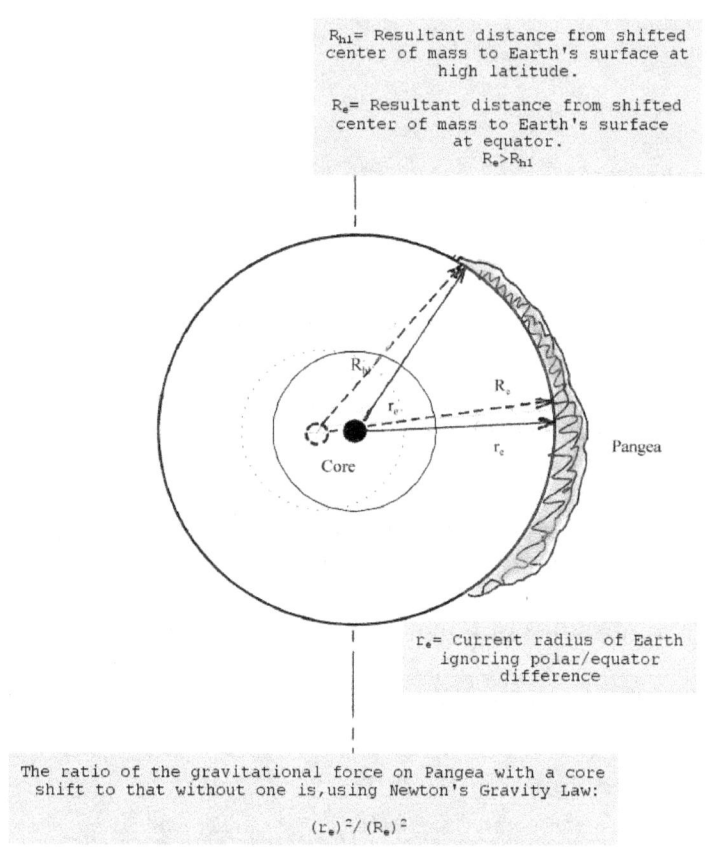

R_{hl}= Resultant distance from shifted center of mass to Earth's surface at high latitude.

R_e= Resultant distance from shifted center of mass to Earth's surface at equator.
$R_e > R_{hl}$

r_e= Current radius of Earth ignoring polar/equator difference

The ratio of the gravitational force on Pangea with a core shift to that without one is, using Newton's Gravity Law:

$$(r_e)^2/(R_e)^2$$

FIG. 10-1 Shift in the inner core(s)

Using a center of mass shift of 1000km would result in a ratio of about .75 (i.e., the weight of an object with the core-shift would be 75% of that without it at the equator) ignoring other factors. Again, this is only a crude estimate because the assumption being made is

that the Earth's mass is all concentrated at a single point. A shift of 1000 km of the center of mass would represent a distance less than half of the inner core's diameter.
Note that:

Inner core diameter = 2400km

Outer core diameter =7000km

A shift of the inner core and outer core would reduce the net surface gravity of Pangea. A core-shift would also help to explain another apparent anomaly described in the following section.

IV. SAUROPOD HABITATS

When studying sauropods, one is struck by a certain pattern which is hard to explain. The relevant literature states that dinosaurs inhabited every continent. Yet when one studies sauropods, there seems to be a high occurrence of sauropods in areas which were in lower latitudes (i.e., closer to the equator) during the Mesozoic Era. The literature also states that during the Mesozoic, tropical conditions existed on all land masses. South America, Africa, mid-western United States, mid to southern Europe, India and southern China are places where abundant sauropod fossils have been found. Places like Canada, northern Europe and Asia, Greenland, Siberia and Antarctica seem to have a dearth or even complete absence of sauropods although other dinosaur remains have been found in those locations.

Is it possible that insufficient searching of those areas is the reason? Is it possible that conditions in those other areas were not conducive to preserving their remains? If that is not the case and there is no other reasonable explanation for their absence in the higher latitudes, then this would support the core-shift explanation embodied in the theory presented in this document.

THE GRAVITY THEORY OF MASS EXTINCTION

It can be seen from Figure 10-1 that the distance to the equator, with the core-shift, is greater than the distance to both of the high (north and south) latitudes on Pangea and therefore the lowest gravitational values would be on land masses closest to the equator.

Figure 10-2 is a representation of Pangea during the late Jurassic Period. The small circles drawn on that map represent locations where sauropod remains have been found including prosauropods. It is only a partial list of sauropods but it serves to illustrate the point being made. The sauropods represented by the small circles are listed on the page following Figure 10-2.

FIG. 10-2 Location of Known Sauropod Remains

SAUROPODS (and prosauropods) USED IN FIGURE 10-2

Name	Location
Isanosaurus	SE Asia (Thailand)
Antetonitrus	South Africa
Euskelosaurus	South Africa (Lesotho,Zimbabwe)
Blikanasaurus	South Africa (Lesotho)
Plateosaurus	Europe (France,Germany, Switzerland)
Lessemsaurus	South America (Argentina)
Riojasaurus	South America (Argentina)
Camelotia	Europe (England)
Melanorosaurus	South Africa
Amargasaurus	South America (Argentina)
Nigersaurus	Africa (Niger)
Cedarosaurus	USA (Utah)
Sauroposeidon	USA (Oklahoma)

THE GRAVITY THEORY OF MASS EXTINCTION

Malawisaurus	South Africa (Malawi)
Agustinia	South America (Argentina)
Phuwiangosaurus	SE Asia (Thailand)
Chubutisaurus	South America (Argentina)
Saltasaurus	South America (Argentina)
Rapetosaurus	Africa (Madagascar)
Jingshanosaurus	China (Lufeng Province)
Yunnanosaurus	China
Massospondylus	South Africa (Lesotho,Namibia,Zimbabwe)
Kunmingosaurus	China (Yunnan)
Kotasaurus	India
Vulcanodon	South Africa (Zimbabwe)
Barapasaurus	India
Omeisaurus	China
Shunosaurus	China
Mamenchisaurus	China, Mongolia
Datousaurus	China
Cetiosaurus	Europe (England, Portugal),Morocco
Amygdalodon	South America (Argentina)

Patagosaurus	South America (Argentina)
Giraffatitan	South Africa (Tanzania)
Lusotitan	Europe (Portugal)
Camarasaurus	USA (N. Mexico to Montana)
Diplodocus	USA (Colorado, Utah, Wyoming)
Supersaurus	USA (Colorado)
Seismosaurus	USA (N. Mexico)
Barosaurus	USA (S. Dakota, Utah), Tanzania
Apatosaurus	USA(Colorado,Wyoming,Utah,Oklahoma)
Brachytrachelopan	South America (Argentina)
Eobrontosaurus	USA (Wyoming)
Dicraeosaurus	South Africa (Tanzania)
Turiasaurus riodevensis	Europe (Spain)
Erketu ellisoni	Mongolia (Gobi desert)
Suuwassea emilieae	USA (Montana)

V. CONCLUSION

A possible explanation for an additional, and probably more important, source of a reduction in the Mesozoic gravitational force on Pangea could be due to a shift of the Earth's solid inner core and the liquid outer core. The breakup of Pangea and drifting apart of the continents would have been accompanied by a shift of the core(s) to a more Earth-centric position, thus increasing the terrestrial gravitational force.

The abundance of sauropod fossils in lower latitudes (i.e., closer to the equator) and their scarcity or absence at higher latitudes would be explainable by a shift in the Earth's core(s) prior to the Mesozoic Era.

CHAPTER 11: THE P-T EXTINCTION- A NEW EXPLANATION

What follows is Part 4 of 4 of original theory. It was published in 2007 on the Internet.

I. INTRODUCTION

The author of this document has written several documents entitled *"The Rise and Fall of The Dinosaurs- The Gravity Theory"* in which the role of a gradually changing gravitational field is posited as the cause for dinosaur gigantism and the primary force behind the eventual extinction of the dinosauria. This happened during the end of the Mesozoic Era (of about 250-65mya). According to that theory, the breaking up and dispersal of the component continents of the super-continent Pangea resulted in a gradually increasing gravitational field, thereby eliminating what the author calls Reduced Gravity Growth (RGG) life forms, resulting in the disruption of dinosaur/mammal equilibrium and allowing the eventual displacement of remaining dinosaurs by mammals.

The End-Permian, also known as Permian-Triassic (P-T) extinction occurred during a period of approximately 251 million years ago (mya). It is considered to be the most devastating period of global extinction in the last 400 million years. An estimated 95% of marine biota and 75% of terrestrial biota disappeared during this period.

Could a changing gravitational field have played a part in this extinction also? The author believes it could have, but not in the same way as the end-Mesozoic extinctions. This document will attempt to explain how gravitational changes could be involved in the Permian-Triassic extinctions. Whether or not there is an underlying gravitational cause for many other extinction periods is a subject that should be investigated. It is very possible that the common causes thought to be responsible for many extinctions were initiated by gravitational change.

THE GRAVITY THEORY OF MASS EXTINCTION

II. P/T EXTINCTION THEORIES

There have been over a half-dozen theories written to explain the P-T extinction. The most well known are summarized below.

1. BOLIDE IMPACT

Due to the widespread belief that an asteroid/comet caused the extinction near the time of the transition between the Cretaceous and Tertiary (K-T) Periods about 65mya, which the author of the theory described in this and prior mentioned documents disagrees with, extinction theorists have made an effort to explain the P-T and other extinction periods based on an extraterrestrial impact.

The evidence to support this P-T extinction theory is almost non-existent. No impact crater of appropriate age and size has been found. There have been no eustatic (i.e., global) indications of a massive impact such as an iridium spike, shocked quartz, etc. This puts this theory near the bottom of the list of believable P-T extinction theories.

2. SEA-LEVEL CHANGE

If one views a graph of the eustatic sea-level fluctuations during the Phanerozoic eon (consisting of the last 542 my spanning the Paleozoic, Mesozoic and Cenozoic Eras) such as the Hallam et al. sea level curve, numerous marine transgressions and regressions (rises and falls in sea levels) occur. Some of them occurred about the time ascribed to extinctions.

There are several reasons for marine transgression/regression. The formation and subsequent melting of ice caps and glaciers causes a regression/transgression sequence. No known glaciation occurred during the P-T transition.

Another cause is the formation of submarine, volcanic ranges, which would cause transgression. The subsequent subsidence of the submerged volcanic ranges would result in regression. Since the ocean floor over 200 million years old has been removed through plate tectonic subduction, the remnants of those marine volcanoes, if they formed during the P-T transition, would not be observable today.

Periods of intense sea-floor spreading causes a rise in sea-level because the less dense lithosphere rising from the spreading centers displace a greater volume of sea water onto the continental margins. Intense sea-floor spreading occurs when there is a lot of movement of the continental and oceanic plates. During the P-T transition, there was reduced plate movement because Pangea was nearing its final stages of consolidation.

The sea-level theorists have posited that a drop of sea levels would have removed the available habitats of the epicontinental seas, thereby causing the extinction of many marine species. There doesn't seem to be an explanation for the high terrestrial extinctions. Others have countered that the extinctions occurred during transgression (i.e., rising sea levels) and there was a flow of anoxic (lacking oxygen) water into the epicontinental seas. Recently, it has been confirmed that transgression occurred during the P-T transition. Based on the reasons given for regression/transgression above, it would be reasonable to assume that submarine volcanism was the cause of rising sea levels at the P-T boundary. There was also massive mantle-plume volcanism occurring during that period at the Siberian Traps.

It has also been pointed out that during the recent Ice Age, when sea-level fell and rose substantially and quickly due to the formation of polar ice, there were insignificant marine and terrestrial extinctions. It would seem logical to infer that if changing sea levels were the primary cause of the P-T extinctions, the ratio of marine/terrestrial extinction would not be 95%/75% but a much higher one, such as 95%/10% and that deep water organisms would be less affected than those in the epicontinental seas.

It is more likely that the coincidence of major fluctuations in sea-level and extinction periods is an indirect one. There is a relationship between sea-level change and the formation of submarine volcanic ranges, and, there seems to be a closer relationship between mantle-plume volcanism and extinction periods. Therefore, there seems to be an indirect relationship between sea-level change and extinctions, not one of cause-and-effect.

3. VOLCANISM

P-T extinction theorists have pondered the coincidence of major extinction periods and eruptions of deep mantle-plume volcanoes. These eruptions are believed to originate at the Earth's outer-core/mantle boundary. They produce what are called Continental Flood Basalts (CFBs) or Large Igneous Provinces (LIPs) when they breach the surface of the Earth. The Deccan Traps in India erupted near the time of the K-T transition and the Siberian Traps in Russia erupted near the time of the P-T transition. These two are the most widely known although there are others. Some of these from the last 251 myrs are shown in Table 11-1 below.

LOCATION	AGE	VOLUME (Million Cubic Km)	PALEO LAT.	DUR-ATION (Myr)
Siberian	248-250	>2	45N	~ 1
Newark	200-202	>1	30N	~ 0.6
Karoo	182-184	>2	45S	0.5 - 1
Antarctica	175-177	>0.5	50S	~ 1
Serra Geral	131-133	>1	40S	??
Rajmahal	115-117	??	50S	~ 2
Madagascar	87-89	??	45S	~ 6
Deccan	65-67	>2	20S	~ 1
North Atlantic	56-58	>1	65N	~ 1
Ethiopia	30-32	~ 1	10N	~ 1
Columbia River	15-17	.25	45N	~1

Table 11-1 Statistics On Flood Basalt Volcanism

Note: these are approximations and different sources will provide somewhat different statistics.

Some theorists have attempted to attribute a causal relationship between mantle-plume volcanism and major extinctions which occurred around the time of those eruptions. While it is conceivable that volcanic activity could cause extinctions if enough material could be injected into the atmosphere in a short period of time, deep mantle volcanism appears to occur over a long period of time, often several million years and is intermittent. The result is a layer cake buildup of lava flows.

The pyroclastic (i.e., explosive type) volcanism such as Mount Saint Helens in Washington, Mount Pinatubo in the Philippines, Mount Pelee, Martinique and Vesuvius in Italy have had an immediate and detrimental effect on the environment due to the rapid release of toxic material. Mantle-plume volcanism, which probably has brief periods of pyroclastic activity, appears to be a more benign, slow acting phenomenon with a less devastating effect on the environment.

The Hawaiian Islands have been, over the last five and a half million years, formed by a hot-spot volcano of the mantle-plume variety. As the underlying Pacific Plate moves northwest, the mantle-plume, being in a fixed position, has formed the Hawaiian Island chain and prior to that the Emperor Seamounts to the northwest of the Hawaiian Islands.

The Siberian Traps volcanism was on a much larger scale than that of the Hawaiian Island volcanism but there has to be a measure of doubt as to whether the former eruptions could directly cause an extinction of the magnitude of the one over the P-T transition. The slow and intermittent particle outgassing of mantle-plume volcanoes would be removed from the atmosphere by rainfall. The carbon dioxide released could contribute to global warming depending on the rate of outgassing. The Traps took place in a relatively high paleolatitude and some scientists claim the extinctions began before the eruptions. If the Traps were the primary contributor to the extinction, one would expect to see the most devastating effects of the extinction in the high northern paleolatitudes compared to other areas but that doesn't seem to be true. Therefore, the question that has to be asked is:

Was the Siberian Traps volcanism near the P-T transition a coincidence or was the volcanism a result of a more dominant force at work?

4. CLIMATE CHANGE

The climate of Pangea did change from a moist one with moderate temperature to an arid, hotter one during the P-T transition. Permian flora that thrived in cooler conditions such as glossopteris, became extinct or near-extinct during this period. A worldwide fungal spike has been found indicating a major destruction of flora in a geologically short period of time. However, it is not known whether the spike is a direct result of the climate change, acid rain, exposure to excessive ultraviolet radiation resulting from the ozone depletion of the stratosphere, deprivation of carbon dioxide by a release of methane or some other reason.

Whether the climate change, by itself, could cause the 95%/75% devastation is a subject of debate. A temperature rise of 6 degrees centigrade has been estimated and it seems like this would not be sufficient to explain the magnitude of the extinction in both marine and terrestrial environments. The reason for the global warming starting near the P-T boundary will probably lead to the primary cause of the destruction.

The global warming across the P-T transition is attributed to two possible causes. They are, the thermal effects of the Siberian Traps volcanism and the release of massive amount of methane from the bottom of the sea. Climate change seems to be one of the several killing mechanisms rather than the primary cause of the extinctions.

5. THE RELEASE OF METHANE FROM THE DEPTHS OF THE SEA/OCEAN

This is the most likely cause of the extinctions at the P-T transition. Methane, contained in methane hydrates at the bottom of the sea/ocean is held in a solid crystalline state by the combination of low temperature and high pressure of the overlying water.

THE GRAVITY THEORY OF MASS EXTINCTION

By studying what scientists call the "carbon isotope change", which was massive during the P-T transition, it has been widely accepted that the only explanation for the magnitude of this change is the disassociation of methane from these hydrates. Carbon isotope analysis is a complex subject but it can be summarized as follows:

There are three naturally occurring isotopes of Carbon: Carbon-12, Carbon-13 and Carbon-14. Carbon-14 is a radioactive isotope which is used to date material up to 100,000 years old but it is not relevant here. Only the non-radioactive Carbon-12 and Carbon-13 are involved.

Scientists study the change in the ratio of C13 to C12 in various environments, both terrestrial and marine. They use a measurement referred to as "delta 13C" to keep track of changes in the ratio. Delta 13C is defined mathematically as:

$$\delta\ ^{13}C = \left(\frac{\left(\frac{^{13}C}{^{12}C}\right)_{sample} - \left(\frac{^{13}C}{^{12}C}\right)_{standard}}{\left(\frac{^{13}C}{^{12}C}\right)_{standard}} \right) * 1000$$

Delta 13C is therefore the percentage change of the ratio of C13 to C12 in a sample material to that in a standard (times 1000). The

92

"Delta" in Delta 13C is often written as the Greek letter δ. In mathematical notation, that letter denotes a change in some quantity.

It turns out that during periods when biotic activity substantially declines, as in extinction periods, the value of delta 13C in the atmosphere and oceans shows a large negative increase. This is called a negative delta 13C spike. The following is a representation of the delta 13C from the late Permian through the early Triassic Periods in a marine location.

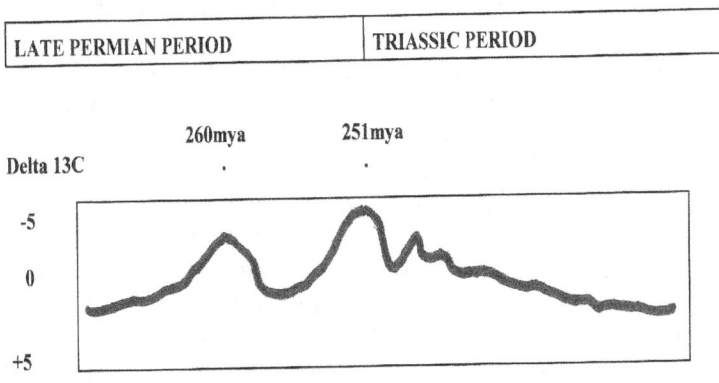

LATE PERMIAN PERIOD	TRIASSIC PERIOD

FIG. 11-1 Delta 13C at a marine location during the late Permian- early Triassic

What is interesting about this graph of the delta 13C change is that there is a small peak in the mid-Permian which corresponds to the end-Guadalupian Stage (about 260mya). This is a period in which there were high rates of extinction prior to the P-T transition extinction. After this peak is a gradual rise that increases more rapidly to a maximum at the P-T boundary (about 251mya). One therefore has to question the significance of the Siberian Traps volcanism as an initiator of the release of the methane hydrates. The Traps volcanism could have exacerbated the negative spike of delta 13C near the P-T boundary though. Also, note the secondary spike, a somewhat smaller peak in the early Triassic. Other representations of delta 13C during this period show multiple peaks well into the Triassic period. There doesn't seem to be a generally accepted explanation for the multiple peaks that extend for several million years into the early Triassic period. An explanation for these peaks will be offered in the next section.

Finally, the release of the methane from the submarine methyl hydrates appears to have had a dual effect relative to the P-T transition extinctions. The more direct effect would have been the anoxia (i.e., lack of oxygen) in the marine environment and terrestrial areas near marine locations. This would cause the asphyxiation of marine and near-marine biota. Methane is a more potent agent of global warming than carbon dioxide and is oxidized in the atmosphere in within 10-50 years to form carbon dioxide. Others suggest that hydrogen sulfide could have been produced, by the release of the methane and its chemical reaction with sulfur compounds, which would also have been toxic to life forms. The indirect effect, after anoxia and/or toxicity, due to the release of methane would be the more gradual but long term extinction caused by severe global warming. Since the rise in the negative delta 13C appears to have started well before the P-T boundary and gradually increased until the rapid rise at the P-T boundary, there has to be doubt as to whether the Siberian Traps was the trigger for the release of the methane. Although, as mentioned before, it could have exacerbated the disassociation of the methane near the boundary. With this in mind, the question that has to be asked is:

What was the trigger for the methane release?

III. THE GRAVITATIONAL TRIGGER

The author of the current theory, as described in this and prior documents describing the Gravitational Theory of Dinosaur Gigantism and Extinction, offers a new and significantly different explanation for the release of the methane from the submarine methyl hydrates.

The thermal effects of the Siberian Traps volcanism has been cited as the trigger for the release of the methane. However, it was noted in the prior section that the negative delta 13C excursion began well before the end of the Permian and extended well into the Triassic Period. Some theorists have suggested that a drop in sea-level, reducing the pressure of the overlying water, was the trigger. Recent study has confirmed that there was a marine transgression at the P-T boundary, thereby eliminating regression as a trigger.

The author of the current theory posits a negative gravitational change on and near the Pangean super-continent as the trigger for the release of methane before, during and after the P-T boundary. As theorized in a prior document entitled *"The Rise And Fall Of The Dinosaurs- Part III, The Gravity Theory of Gigantism and Extinction"*, a shift of the Earth's solid inner core and the liquid outer core, is a function of the movement of the continental plates. The movement of the both cores are linked to these plates by a scientific principle known as the Conservation of Angular Momentum. As the continents coalesced to form Pangea, there was a corresponding shift of the both cores away from an Earth-centric position to a position further away from the center of mass of the Pangean super-continent, thereby lessening the gravitational force on Pangea. Because the Pangean land masses were distributed extensively over a primarily north-south expanse, there was a major latitudinal gravitational gradient and a lesser longitudinal gradient.

Pangea was in the end-stage of its consolidation during the P-T transition. Baltica and the south China Blocks were the only large land masses still being consolidated. The collision of the North and south China Blocks occurred during the Permian-Triassic transition.

In other words, the gravitational force on the surface of Pangea was approaching its lowest levels. And, comparable to a dramatic drop in sea-level, the reduction in gravitational force would result in a gradual but significant drop in pressure at the bottom of the sea/ocean within and near Pangea.

Any movement of a continental plate would cause a corresponding movement of the inner/outer cores. The magnitude and direction of the core movement would be dependent on the mass of the moving plate and its direction of movement. The movement of the solid inner core would generate pressure changes within the molten outer core not unlike a piston moving within a cylinder of water. These pressure changes would be responsible for the mantle-plume volcanic activity. Some have observed the coincidence of plate tectonic activity and major mantle-plume volcanism but have not been able to explain the linkage. The current theory attempts to explain that linkage.

The major instances of large eruptions of deep mantle-plume volcanism have occurred during the period when Pangea was almost fully consolidated in the late Permian until its breakup was in full swing in the end-Cretaceous. Such major eruptions are not known prior to the Permian and those that have occurred since the K-T boundary have gradually decreased in size (i.e., volume of lava produced) as can be seen in Table 11-1. This would be expected because as the continents dispersed, the gravitational component from each continent would be diminished toward the original consolidated Pangean location, thereby lessening its influence on the inner core's movement.

Some have also observed that the major flows of lava from the Flood Basalts that are associated with extinction periods seem to lag the start of the extinction periods thereby implying a non-causal relationship. This would be consistent with the current theory as it applies to the P-T transition. A continental plate movement would invoke a synchronous core movement (due to the linkage previously described), which would invoke a synchronous gravitational change on Pangea, which would invoke a synchronous release of methane (when the inner/outer cores were moving away from Pangea).

However, the inner-core movement would generate a mushroom plume starting at the outer-core/mantle boundary and rising to the surface over a much longer period of time. A lag in extinction/volcanism would be the result.

V. CONCLUSION

1. The P-T extinctions were most likely the direct result of oxygen removal from marine and terrestrial locations along with possible hydrogen sulfide toxicity, and, an indirect result (of global warming) due to the release of methane from submarine methyl hydrates. The release of the vast stores of methane seems to be the only explanation of the large negative swings in the delta 13C values near, during and after the P-T boundary. Disassociation of methane from submarine hydrates can occur with large temperature increases and/or large drops in pressure. The trigger for the release of the methane doesn't appear to be from the Siberian Traps because the methane release, based on delta 13C history, predates the P-T boundary. Thermal effects of the Traps volcanism may have enhanced the release of methane near the boundary.

2. The disassociation of the methyl hydrates would probably have occurred in pulses corresponding to the delta 13C graph shown in Figure 11-1 and would have continued well into the Triassic Period. This would help to explain the unusually long recovery period.

3. The theory embodied in this document does not posit a new direct "killing mechanism" to explain the extinctions. What it posits is a gravitational trigger for the release of methane from submarine methyl hydrates before, during and after the P-T boundary. The lowest values of gravitational forces on Pangea occurred when the final land masses were coalescing in the late Permian/early Triassic Periods thereby lowering the pressure on the submarine hydrates.

THE GRAVITY THEORY OF MASS EXTINCTION

4. A gravitational linkage between continental plate movement/shifting of the Earth's inner and outer cores/ fluctuations of surface gravity on Pangea, have been described and the author believes that this was the primary reason for the release of the methane. The pulses in the delta 13C before during and after the P-T boundary, the author believes, correspond to movements in the continental plates which are gravitationally linked to the Earth's inner core as described.

5. The massive mantle-plume volcanism, including the Siberian Traps, is also a result of the gravitational linkage described above. As described in this document, the mantle-plume volcanism is posited to be a result of pressure changes within the molten outer core due to the inner core oscillations. The surface effects of the volcanism would lag its cause (i.e., large tectonic plate movement).

6. Also, the possibility that other extinction periods were initiated by gravitational changes based on the theory described in this and previously mentioned documents must be considered.

TERRESTRIAL EXTINCTIONS ....

MESOZOIC TERRESTRIAL EXTINCTIONS

Age mya

| 200 | . . . . | 150 | . . . . | 100 | . . . . | 50 | . . |

........Jurassic..........^...........Cretaceous................^.Cenozoic

S P T B B C O K T B V H B A A C T C S C M P   E

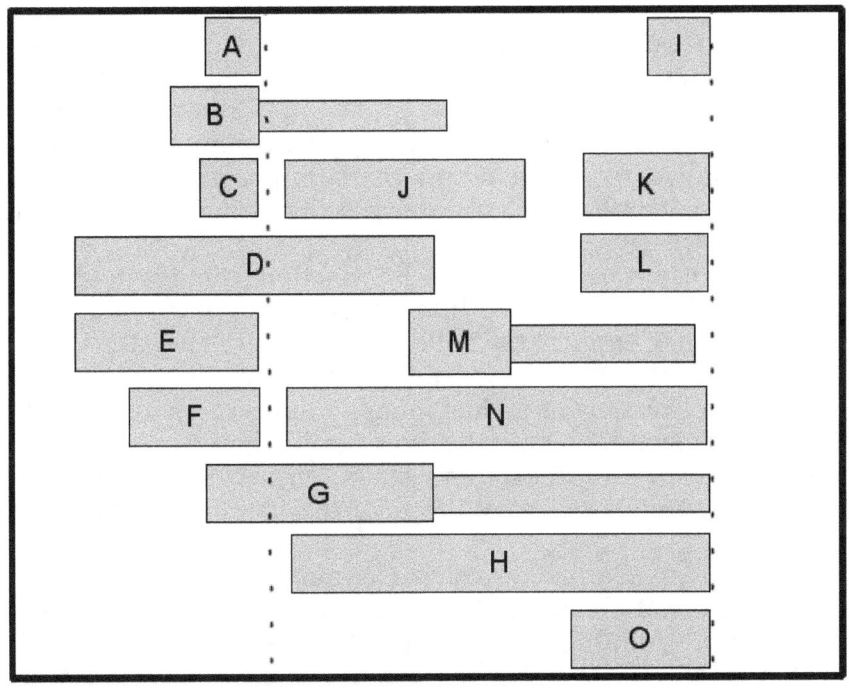

FIG. 12-1 Mesozoic tetrapod extinctions, adapted
from Dr. Robert T. Bakker, 1977

99

THE GRAVITY THEORY OF MASS EXTINCTION

The Mesozoic terrestrial (large herbivores) extinctions shown above are adapted from Dr. Robert T. Bakker's 1977 paper *"Tetrapod Mass Extinctions."* The following terrestrial families are represented:

A=camarasaurids, B=diplodocids, C=stegosaurids, D=brachiosaurids, E=cetiosaurids, F=camptosaurids, G=hypsilophodontids, H=panoplosaurids, I=ceratopsids, J=iguanodontids, K=hadrosaurids, L=protoceratopsids, M=titanosaurids, N=pachycephalosaurids, O=euoplocephalids

The above chart of tetrapod extinctions in Figure 12-1 raises an important question. There is unquestionably a mass extinction at both the Jurassic/Cretaceous and the Cretaceous/Tertiary boundaries. If we do not find bolide impact nor massive flood basalt volcanism at one of those two periods, is it possible that both extinctions are not the result of either one of those causes but are the result of a third cause?

Frequent references to the K-T extinctions often state that the dinosaurs "suddenly went extinct." This terminology often sways public opinion to support the bolide impact cause for extinction. In reality, most of the familiar dinosaur forms, including the sauropods and the stegosaurs, went extinct many millions of years before the Chicxulub impact. And, even the more familiar late forms including triceratops, hadrosaurs and tyrannosaurs appear to have disappeared or became extremely rare by the time the famous clay layer was formed.

The most notable terrestrial animals that became extinct throughout the Mesozoic are the largest ones. Among them are the sauropods, stegosaurs, spinosaurs and the pterosaurs.

If the GTME is correct, the terrestrial extinctions of the largest animals should display a pattern that is predictable and recognizable. That pattern would be an evolutionary increase in size of selective life-forms of small land vertebrates to gigantic forms, relative to modern times, followed by their downsizing and eventual extinction.

The following is a summary of the most familiar terrestrial animals that are relevant to the current theory.

Sauropod

Stegosaur

Pterosaur

Spinosaur

Ankylosaur

FIG. 12-2 Examples of large extinct terrestrial animals

THE GRAVITY THEORY OF MASS EXTINCTION

SAUROPODS..............................

The earliest sauropods, the prosauropods, appeared in the late Triassic and consist of small members such as Plateosaurus, Blikanasaurus and Melanorosaurus. Most prosauropods are found in the Norian Age (216-203 mya) of the late Triassic.

The prosauropods were replaced by much larger sauropods. The most familiar of these descendants are the Brachiosaurids and Diplodocids. The Brachiosaurids are referred to as high-browsers because their long necks are much more vertically oriented than other sauropods. They also have forelimbs that are considerably longer than their hindlimbs, probably an evolutionary change to aid in high browsing.

The Diplodocids were contemporaneous with the Brachiosaurids. An abundance of fossils of both sauropods have been found in the western United States. Unlike the Brachiosaurids, the Diplodocids had forelimbs that were much shorter than their rear limbs. Their neck and tail were extremely long and were probably held more horizontally than vertically. It is believed they were able to feed tripodally and at a much wider vertical range than Diplodocids.

The late Diplodocids and Brachiosaurids, which became extinct in the early Cretaceous, had become smaller and developed backbone spines. Amargasaurus and Nigersaurus are two examples. Sauroposeidon, the last of the Brachiosaurids in North America, seems to have bucked the trend of a diminution in size. However, the neck bone structure contains air cells for weight reduction indicating a slimmer animal than a brachiosaur. Artistic representations of Sauroposeidon indicate a much smaller tail, which would confirm a lighter head/neck structure.

The Titanosaurids were the last of the large sauropods. In the *Titanosaur Dilemma* section of *Chapter 9*, the reasons why titanosaurs would be able to defy an increasing gravitational field were given, they are:

102

1. Much wider hips and leg stance allowing more of the body mass to be closer to the ground.

2. Not a migratory animal as were most earlier sauropods, thereby requiring less energy expenditure for a large animal.

3. Shorter necks reduced need for high blood pressure to supply blood to brain.

4. It has been found that the hips are fused to 6 vertebrae instead of the usual 5 for sauropods thereby providing more weight bearing strength.

5. Body armor is very common in many titanosaurs. This could be an evolutionary attempt to offset their smaller size with a defensive mechanism to fend off predators.

STEGOSAURS...................................

Stegosaurus were massive, quadrupedal, herbivorous dinosaurs that are familiar to most people. They were designated as the state fossil of Colorado,USA. Physically, they have the following characteristics:

- Short forelimbs compared to rear limbs.
- Arched back.
- Head positioned near ground.
- Stiffened tail with spikes held high above ground.
- Numerous plates from neck to tail along spine.

THE GRAVITY THEORY OF MASS EXTINCTION

The spikes were obviously for defensive purposes and it seems logical to assume that the elevated posterior enabled the stegosaurians to see approaching predators by looking below their body since their head was positioned low for feeding. There seems to be two explanations for the plates. Defense is the most probable explanation but thermoregulation, the ability to absorb and dissipate heat, is the alternate explanation. The excessive armor was necessary because the large theropods, Allosaurus and Ceratosaurus were contemporaneous with the stegosaurians.

Many primitive stegosaurians fossils have been found in China and it is believed they originated in that region. Stegosaurus is the most familiar stegosaurian and was probably the largest, averaging around 9 meters (30 ft) long and 4 meters (14 ft) tall. The hind feet each had three short toes. The forelimbs were much shorter than the more massive hindlimbs, which resulted in an unusual arched posture. The tail, with two pairs of spikes, were held in a high position, undoubtedly positioned to defend against the large theropod predators. The head of Stegosaurus was positioned relatively low down, probably no higher than 1 meter above the ground permitting browsing for low level plant life and facilitating its awareness of predators approaching from the rear.

Most stegosaurian fossils have been found in upper Jurassic deposits. These four Ages (Stages) of the Jurassic are:

- Callovian (164.7 - 161.2 mya)...The earliest and most primitive forms
- Oxfordian (161.2 - 155.7 mya)
- Kimmeridgian (155.7 - 150.8 mya)
- Tithonian (150.8 - 145.5 mya)

Several primitive forms predate the range given above, including Huayangosaurus (from China) and Scelidosaurus (from England). The stegosaurians became extinct at the Jurassic/Cretaceous boundary. No definitive explanation for their demise has been given.

ANKYLOSAURS ....

The ankylosaurians were herbivorous, quadrupedal dinosaurs that arose in the early Jurassic period in China and their fossils have been found on every continent except Africa. They were smaller than the stegosaurians but were heavily armored also, primarily with bony scutes. They did not have the disproportionately shorter forelimbs that the stegosaurians had nor did they have the tail spikes.

The ankylosauria consists of two families, the nodosaurids and the ankylosaurids. The latter had bony clubs at the end of their tails (except for polacanthids) but the nodosaurids did not. Most nodosaurid fossils are from the Campanian Age (83.5 mya) through the Maastrichtian Age (65.5 mya). They are the more primitive group, having narrow heads and large spikes protruding from the shoulder area pointing forward and downward. They also had broad backs covered with armor. Edmontonia, from the western United States, Canada and Alaska is a good example of the nodosaurids.

The ankylosaurids that proliferated in the Cretaceous are believed to have evolved from the thyreophorans of the mid-Jurassic Period. The ankylosaurids had wider bodies and thicker armor than the nodosaurids. The last of the ankylosaurids included Pinacosaurus, Talarurus and Euoplocephalus. The ankylosaurids did not survive beyond the Cretaceous.

> The earlier extinction of the larger stegosaurids (at the Jurassic/Cretaceous boundary and the much later extinction of ankylosaurids, a similar but smaller animal that probably evolved from the same ancestor, the thyreophorans) is consistent with the GTME).

SPINOSAURS.................................

Spinosaurus, which means "spine lizard" was an extremely large theropod dinosaur from the mid-Cretaceous period of about 100 mya to 93 mya. Their fossils have been found along the northern coast of

105

Africa which was the western part of the Tethys Sea during the Mesozoic Era. Their name is based on the large dorsal sail-like structure which was at least 6 feet in height. There are several opinions about the function of that structure including thermoregulation, sexual display and intimidation display.

Spinosaurs are believed to have been the largest of the carnivorous theropods, at least in length and height. Various estimates put their length at 16 to 18 meters (about 52 feet to 60 feet) and their weight from 7 to at least, if not higher than, 10 tons. A recent paper gives an estimate of 12 to 14 meters for length and 12 to 21 tons for weight. They were larger than the more compact Tyrannosaurus rex. Their physical structure and the location of their fossils indicate they were primarily fish eaters although fossilized stomach remains contain pterosaur and iguanodon bones.

PTEROSAURS ....
Pterosaurs are considered to be the first vertebrates to have developed flight. They are categorized as reptiles rather than dinosaurs although recent thinking is that they were warm-blooded, which casts some doubt whether they are classified correctly. Their flying behavior, especially for smaller species, was very similar to birds and their high energy mode of locomotion mandates a warm-blooded physiology. As with extant birds, their bones were hollow and they had a keeled breastbone for the attachment of muscles to power wings. The wings were attached to the evolved, lengthened fourth finger of the arms but there is some controversy about whether the wings were attached to the hindlimbs. Actinofibrillae, which are closely woven fibers, provided strength and durability to the wings. It is believed that most, if not all, pterosaurs had webbed feet.

Pterosaurs are divided into two broad sub-classes:

Rhamphorhynchoidea

The Rhamphorhynchoid pterosaurs are from the Oxfordian to the Kimmeridgian Ages (about 161.2 mya to 150.8 mya) of the late Jurassic and disappeared at the Jurassic/Cretaceous boundary. These pterosaurs were the smallest and most had long tails. Examples are Batrachognathus, Jeholopterus, Rhamphorhynchus and Sordes. The discovery of the presence of hair on a Sordes fossil convinced paleontologists that pterosaurs were warm-blooded.

Pterodactyloidea

The Pterodactyloid pterosaurs evolved from, and replaced, the Rhamphorhynchoid pterosaurs. They appeared in the early Cretaceous and the group eventually became extinct toward the end of that period. They were larger than their predecessors, had shorter tails and longer necks. They also developed much longer wing metacarpals giving them long wingspans. Some examples are Ctenochasma, Gallodactylus, Gnathosaurus and Pterodactylus.

In the 1970s, the fossil remains of one of the largest pterosaurs, Quetzalcoatlus northropi, was uncovered in the Big Bend area of Texas, USA. The initial estimate of its wingspan was 20 meters (63 feet). When this wingspan was publicized, aeronautical engineers protested that it would be impossible for that pterosaur to have flown. Under pressure, the paleontologists who made the discovery revised their estimate down to near 40 feet. The aeronautical engineers, of course, did not consider the possibility that surface gravity might have been lower 65 to 70 mya than it is today.

Few reasons for the eventual extinction of the pterosaurs at the end of the Cretaceous Period have been offered. Since birds survived the end-Cretaceous extinctions, the question of why pterosaurs, that "made a living" in a similar way, would become extinct is unanswered. Since the fossil record clearly indicates that the pterosaurs were gradually becoming larger, one answer proposed was that they were being outcompeted by birds. If that were true, the pterosaurs would more likely have evolved smaller, rather than larger, body sizes. Another opinion is they were primarily ocean-going fishers and that the "K-T mass extinction" affected their food supply.

THE GRAVITY THEORY OF MASS EXTINCTION
MARINE EXTINCTIONS ....
The following chart identifies many of the marine vertebrates that
became extinct during the Mesozoic Era.

MESOZOIC MARINE EXTINCTIONS

Age mya 200 . . . . 150 . . . . 100 . . . . 50 . .
Triassic...^...Jurassic............^.....Cretaceous.....^..Cenozoic
SA L C N R H S P T B B C O T B V HB A AC TC SM P EO M

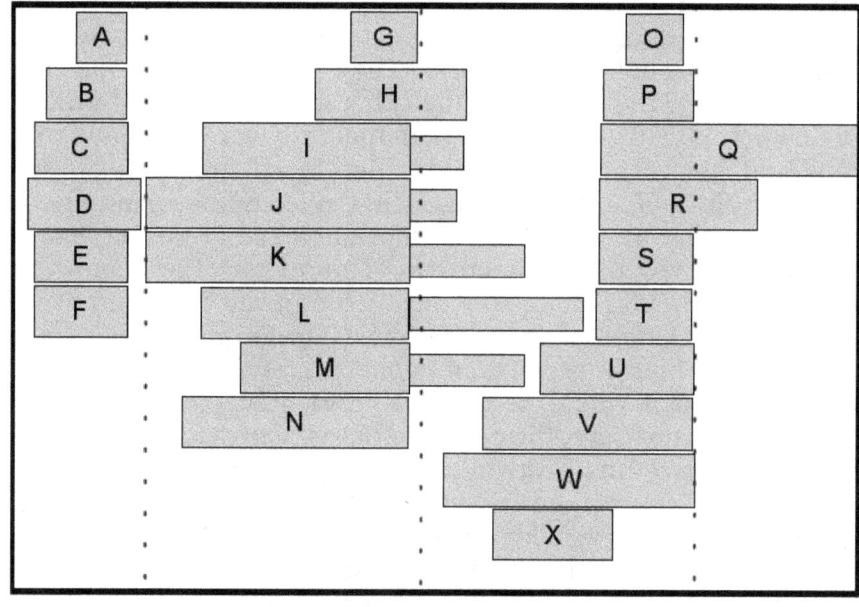

FIG. 12-3 Mesozoic marine extinctions,
   adapted from Dr. Robert T. Bakker, 1977

The Mesozoic marine extinctions shown above are adapted from Dr. Robert T. Bakker's 1977 paper "Tetrapod Mass Extinctions. The following marine families are represented:

A=henodontids, B=pachypleurosaurids, C=mixosaurids, D=placocheliids, E=nothosaurids, F=shastasaurids, G=pliosaurids, H=metriorhynchids, I=rhomaleosaurids, J=ichthyosaurids, K=plesiosaurids and cryptoclidids, L=stenopterygiids, M=teleosaurids, N=rhamphorhynchids, O=protostegids, P=mosasaurids, Q=cheloniids, R=toxocheliids, S=ichthyornids, T=hesperornids, U=elasmosaurids, V=polycotylids, W=ornithocheirids, X=ornithodesmids

The following are some of the well known marine animals (vertebrates and invertebrates) that became extinct in the later part of the Mesozoic Era.

INOCERAMIDS.................................

The inoceramids were clams (i.e., bivalves) that were as large as six feet in diameter. They lived on the sea floor of the ocean and had been in existence for approximately 200 million years. Many inoceramid fossils have been found in the areas where the Western Interior Seaway had flooded the USA during the Cretaceous. They had a global distribution during that time.

The inoceramids had a thick shell that had a pearly luster. Their shells had prominent concentric, semicircular growth rings. Many paleontologists believe that the giant size of these clams was attributable to the need to filter large quantities of water in the murky, oxygen deficient depths where they lived.

Peter Douglas Ward of the University of Washington and Charles Marshall of the University of California studied molluscan extinctions in the western European Tethys region. In the Bay of Biscay region of Zumaya, Spain, during a time that represents 3-4 million years of the latest Cretaceous Period they found fossils from

7 species of inoceramids. Using mathematical techniques (known as rarefaction analysis) to estimate times of extinction, they estimated that 6 of the 7 inoceramid species went extinct well before the K-T boundary. The sole surviving species was the small enigmatic form Tenuipteria. It appeared to have continued to within 1.5 meters of the K-T boundary.

The study of inoceramid fossils in northern Japan (at Hokkaido) indicates a major extinction of inoceramids at the C/T (Cenomanian/Turonian) boundary, which is about 93.5 mya. According to that study, all late Cenomanian species were replaced by early Turonian species in that region. What was noted about the new inoceramid species by Akinori Takashi is:

"This change in generic composition, accompanied by stunting, a decrease in interspecific size variation, and predominance of cosmopolitan species evidently occurred immediately after the C/T transition."

The study concludes that the results mentioned above were caused by an oxygen depletion condition associated with Oceanic Anoxic Event 2.

Kenneth G. McLeod, of the Univ. Of Washington, believes that the inoceramid extinctions were caused by a cooling of the Earth during the Maastrichtian. He reasoned that the drop in temperature caused a larger pole/equator differential in water temperature resulting in a substantial inflow of oxygen into the deep water habitat of the inoceramids. While the increase in oxygen in the deep habitats of the inoceramids during the Maastrichtian might have affected them negatively, one has to question whether it could completely annihilate them. And, how would the earlier extinctions at the Cenomanian/Turonian boundary be explained?

The GTME would explain the extinction of the inoceramids as a series of increasing gravitational pulses, one being at or near the Cenomanian/Turonian boundary and another closer to the Mid-Maastrichtian. The transition to smaller inoceramid forms at the C/T boundary would be consistent with rapidly increasing surface gravity at that time.

MOSASAURS.....................................

The mosasaurs were considered the top marine predator of the last 20 million years of the Cretaceous Period. From the Turonian Age (93.5mya) to the end of the Maastrichtian Age (65.5 mya) these reptile predators were found worldwide. Like the ichthyosaurs that preceded them, they gave birth to live young (i.e., viviparity). The mosasaurs are considered to be related to snakes or monitor lizards based on skull and jaw similarities.

There was a wide diversity in the size of mosasaurs. Carinodens belgicus was the smallest at 3.5 meters while Hainosaurus at almost 18 meters is the largest one known. They evolved a long and streamlined body type and were strong swimmers. Their broad tail provided their locomotive power similar to the way an eel propels its way in the water with a side to side undulating motion. Their double hinged jaw allowed them to swallow their prey whole and they were more likely ambush rather than pursuit predators.

Mosasaur fossils are found extensively in the USA in areas formerly inundated by the great Western Interior Seaway. Their fossils have been found in Europe, Africa, Antarctica, South America, New Zealand and Canada. The discovery of the mosasaur fossils in

111

THE GRAVITY THEORY OF MASS EXTINCTION

the Maastrichtian limestone beds of the Netherlands is the reason why the final six million year epoch of the Cretaceous is known as the Maastrichtian.

Mosasaur taxa, such as Clidastes and Halisaurus underwent an abrupt extinction or series of extinctions near the Early/Late Campanian (~72 mya) boundary (per Lindgren, Siverson 2003) in, at least, Europe and North America. An explanation for this extinction has not been given. See page 65 for a drawing of a mosasaur.

According to Bardet (1994), pliosaurs and protostegid turtles become very scarce during the Maastrichtian prior to their final extinction at the end of that period. The protostegid marine turtles were the largest turtles. For example, Archelon was the size of a Volkswagon beetle.

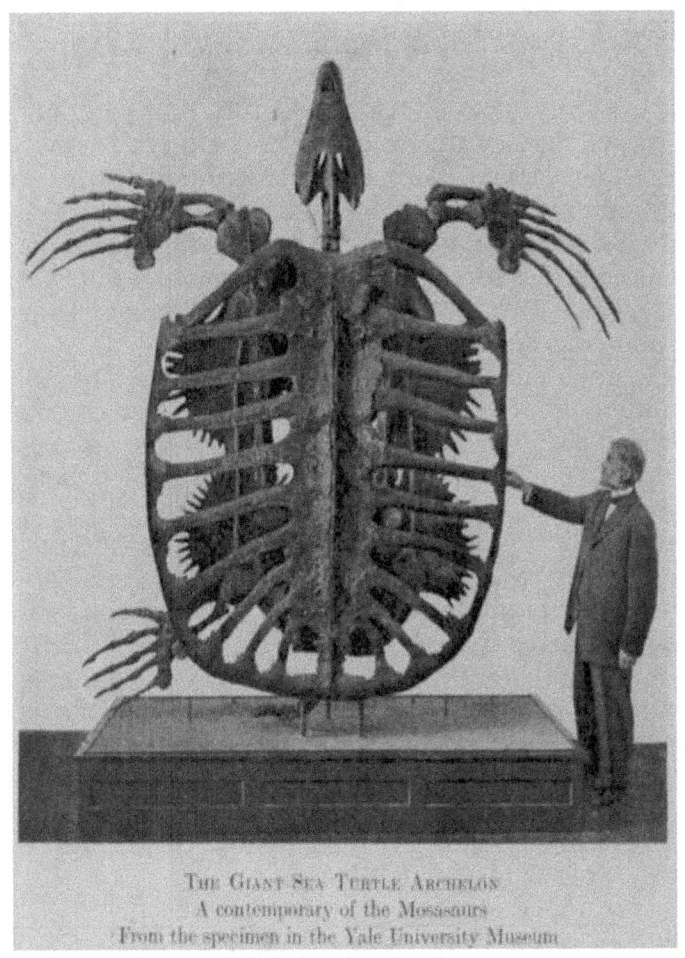

THE GIANT SEA TURTLE ARCHELON
A contemporary of the Mosasaurs
From the specimen in the Yale University Museum

FIG. 12-4 Archelon
Courtesy Yale University Museum

THE GRAVITY THEORY OF MASS EXTINCTION

ICHTHYOSAURS.....................................

Ichthyosaurs were swimming reptiles that originated in the early Triassic. The early forms were not similar to the more familiar dolphin-like later forms that most of us are familiar with. The earlier forms lacked the fish-like tail and dorsal fins. They had large thick tails and used the undulating motion of their bodies to propel them in the water. Therefore, they were slower swimmers than their descendants. Early examples of ichthyosaurs include Utatsusaurus, Cymbospondylus and Mixosaurus.

The ichthyosaurs of the Jurassic developed a streamlined form very similar to dolphins. This is an example of convergent evolution whereby completely different life forms, in this case reptiles and mammals, independently evolve similar physical structures. They also gave live birth to their young.

Ichthyosaurs disappeared in the early Berriasian (145 mya) and Valanginian (140 mya) Stages of the early Cretaceous Period. It was originally thought that the decline of ammonites, which was thought to be their primary food supply, was the cause. When the fossil remains of an ichthyosaur was found that contained the remnants of turtles and a sea bird, that explanation for extinction lost favor. Another explanation for their extinction is the competition from plesiosaurs. It is hard to accept the premise that the ichthyosaurus lineage, which began in the early Triassic and were the dominant marine predator could have been displaced by the plesiosaurs. See page 64 for a drawing of an ichthyosaur.

The GTME attributes the extinction of the ichthyosaurs to a pulse of increasing surface gravity in the early Cretaceous. The live-birth characteristic of this reptile, its high speed pursuit strategy and its physical structure were susceptible to higher gravitation.

AMMONITES.....................................

Today, there are no ammonites living in the planet's oceans or seas. Their final extinction was not at the K-T boundary at the end of the Cretaceous Period as is commonly believed but in the subsequent lower Danian Age. This period follows the upper Maastrichtian Age, the last period of the Mesozoic. Evidence of this will be presented later in this chapter. Because they were prevalent worldwide and were constantly evolving new forms, the ammonites have been recognized as 'index fossils'. This permits the dating of the lithified sediments in which the fossilized remains are found.

The ammonites were shelled members of a group of marine organisms called cephalopod mollusks. They were named after the Egyptian god of life and reproduction, Ammon. This class of the Cephalopoda is divided into a total of 7 subclasses, they are:

ACTINOCERATOIDEA
AMMONOIDEA
BACTRITOIDEA
COLEOIDEA
ENDOCERATOIDEA
NAUTILOIDEA
ORTHOCERATOIDEA

Out of the above subclasses, only the Coleoidea (containing squid, cuttlefish and octopus, which didn't appear until the Jurassic) and Nautiloidea (containing the nautilus) exist today. The Ammonoidea became extinct in the early Danian Age of the early Cenozoic Era. All three groups evolved independently.

The belemnites were the first members of the Coleoidea to appear in the Mississippian (about 350 mya) and they also didn't become extinct until the early Cenozoic. The terms "ammonoids and "ammonites" will be used interchangeably as well as the terms "nautiloid" and "nautilus."

Ammonites evolved from the still extant nautiloids, approximately 400 mya. Over 80 families of the Ammonoidea lived in the Triassic Period. Drawings of an ammonite and a nautilus are shown in the figures below.

FIG. 12-5
Ammonite

FIG. 12-6
Nautilus

Note that the above figures represent a single example of these two cephalopods. Among the ammonoids, there were many different shell types including straight, curved, coiled , corkscrew, etc.

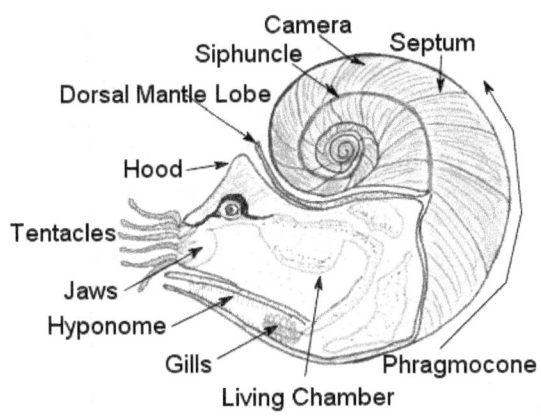

AMMONITE STRUCTURE

Camera
Siphuncle
Septum
Dorsal Mantle Lobe
Hood
Tentacles
Jaws
Hyponome
Gills
Living Chamber
Phragmocone

FIG. 12-7

Both the ammonite and the nautilus have chambered shells. The chambers are called camerae. The live animal occupied the open and largest end of the shell and as they grew, a new chamber was added, being sealed off by a transverse wall called a septum. The connection between the septum and the peripheral shell is called the suture and these patterns have diverse forms. The chambered part of the shell is called the phragmocone. These chambers can be dynamically filled with gas or fluid in order to achieve neutral buoyancy in the water column enabling flotation at any depth. This filling process is accomplished through a tube, called a siphuncle, which passes through each chamber. In the nautilus, this tube passes through the center of the septa while it runs along the ventral periphery of the ammonite. Ammonites can propel themselves by expanding their hyponome (a bellow-like chamber), allowing it to fill with water, and then contracting it to expel a jet stream of water. Tentacles surround

117

the beak-like jaws and are used to grasp and draw in food. The hood protects the open end of the shell from predators.

In general, the nautilus has a thicker shell, enabling it to descend to a depth of about 400 meters while the ammonite was limited to a depth of about half of that. If either of these animals exceeded the depth constrained by their shell construction, they would implode due to the excessive hydrostatic pressure. The corresponding depth is known as the implosion depth.

While the basic structure of the nautilus shell seems to have changed little during the Mesozoic, the opposite is true of the ammonites. Their shell structure evolved a prolific assortment of shapes and embellishments. Ribbing, tuberculations (i.e., small raised rounded nodules), elaborate sutures and other shell characteristics seemed to proliferate and increase especially in the Cretaceous Period. Some paleontologists who have studied this phenomenon claim that the following are evolutionary reasons for the changes:

1. To make the shell less pervious to the newly evolved shell breaking (durophagous) marine predators such as fish and benthic crabs.

2. To strengthen the shell to allow access to greater depth in the water column while adding minimum additional weight. Thickening of the entire shell, as in the nautilus, would permit this but would make them heavier and slower and more vulnerable to other predators.

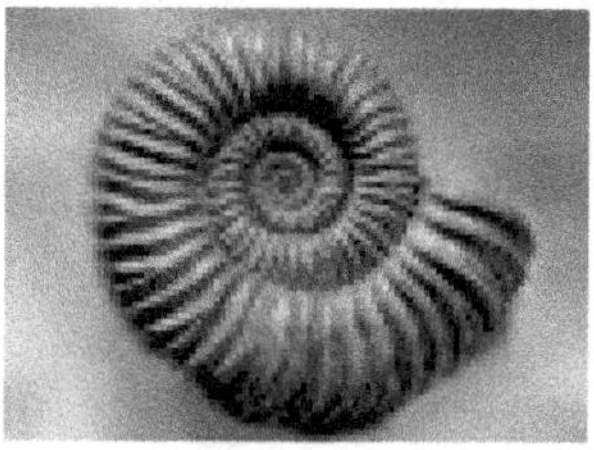

FIG. 12-8 Heavily ribbed Ammonite

Until the K-T transition, both the ammonoids and nautiloids managed to survive all of the prior extinction periods. Although most species were eliminated in each of these periods, the few surviving species were able to recover and radiate to a highly diversified level. The ammonites diversified to a much greater extent. Their shell structures included straight, coiled, uncoiled, curved, cone-shaped, corkscrew and other shapes, and many had complex sutures.

Peter Ward of the University of Washington and Philip Signor of the University of California at Davis studied the fossil remains of ammonites at Zumaya in northern Spain and found that the final demise of the ammonites began 4 to 5 million years before K-T.

If the Gravity Theory of Mass Extinction is correct, the ammonites should follow the same general extinction pattern as dinosaurs. But since we are comparing marine versus terrestrial creatures, the comparison becomes less rigorous. With an increasing strength of gravity at the Earth's surface, it would be expected that the mitigating effect of buoyancy would cause the marine animals to lag their terrestrial counterparts relative to the effects of gravitational changes. Even though the record of the terrestrial dinosaurs fossils is much sparser than that of the ammonites, their demise appears to begin long before that of the ammonites in the late Cretaceous.

During the Mesozoic Era, the diversity in size comparison between terrestrial and marine fauna does compare favorably. Ammonites ranged in size from about one half inch to as much as 7 feet in diameter. And, just as no terrestrial animal has evolved to anywhere near the size of the largest sauropods since K-T time, no external shelled cephalopod, including the nautilus, has come anywhere near the size of the largest ammonite since that time. It must be noted that the squid and the octopus are members of the cephalopod family (subclass Coleoidea) and that raises a problem that has baffled the experts. Since the squid and octopus occupied the same shallow waters at the time the ammonoids went extinct, why did they survive? Today's squid and octopi, when young, feed on zooplankton, food that should have been wiped out, according to pro-impact theorists, by an earth shattering impact and the subsequent acid rain generated. Therefore, the issue of marine food supply reduction as a general cause of K-T extinctions, whether from a reduction in photosynthesis or some other cause, must be questioned. What is significant is that the squid and octopus do not have large exterior shells. The present theory makes the following assertion:

Any shelled organism, especially those where the mobility of that organism would be substantially affected by a (positive) change in hydrostatic pressure with depth, would be more likely to become extinct. The GTME posits a higher gravitational field for that increase in hydrostatic pressure.

Did the ammonites also undergo a drastic reduction in physical size in the upper Maastrichtian Age? Jost Wiedmann of Tubingen University studied Mediterranean ammonites from this period and found that at Zumaya, ammonites appear to have stunted growth. Zumaya is an ancient fishing village in the Basque region of northern Spain. He concluded that ammonite dwarfism was caused by some constriction of their food supply, possibly related to sea-level changes. Although some have claimed that these fossils were those of juvenile ammonites, could the reduced size instead be an evolutionary response to a major pulse in a gravitational increase?

The Zumaya location also drew the attention of Peter Ward, an expert on ammonites mentioned earlier. It is one of a few places on the planet where there is a continuous stratigraphic exposure of marine fossils in the transition from the upper Cretaceous Period through the lower Tertiary Period. The ancient cliffs there were uplifted when the Pyranees were formed. Ward would become directly involved in the Alvarez (Impact) Extinction Theory. After that theory was announced, Ward was invited to the University of California at Berkeley in 1981 to present his views on ammonite extinction to paleontologists. The Alvarezes, both father and son were present. Ward's conclusion, the result of a theoretical study, was that the ammonites had gone extinct suddenly, thereby supporting the impact scenario. The Alvarezes were delighted to hear this and promptly invited Ward to dinner. Another attendee at the presentation was Berkeley paleontologist Dr. William A. Clemens who disagreed with the Alvarez theory. He left without comment.

During the following year Ward, along with Jost Wiedmann, did extensive field work on Cretaceous ammonites in the Boundary Bay area at Zumaya. Although they did find many ammonite fossils from a half dozen species, they could not find any ammonite fossils within 30 feet of the clay layer associated with the K-T transition. They concluded that the ammonites had gradually become extinct long before K-T time. A second presentation was given at Berkeley where their results were presented. This time Clemens invited Ward to Dinner.

In 1987 Ward returned to Zumaya and traveled to locations within 100 miles of his prior ammonite hunting grounds. These included Hendaye, a town on the Spanish-French border and Biarritz about 10 miles away. Both locations had numerous ammonite fossils that were very close to the K-T boundary (i.e., within one meter). Ward came to the conclusion that the reason he found no ammonites within 30 feet of the boundary at the original site at Boundary Bay, Zumaya, was that at the end of the Cretaceous, Zumaya was in the deepest part of the basin. He wrote:

"..at the end of the Cretaceous, Zumaya was in the deepest part of the basin, at depths too great to sustain many ammonites."

The implication is that most, if not all, of the ammonites in that region needed to access the bottom of the basin for food, refuge from predators or some other vital reason. The depth factor sounds like a

121

good interpretation for the absence of the ammonite fossils in the last 30 feet while there are abundant fossils below that. This could be a reasonable explanation because there was a major regression during the late Maastrichtian Age (allowing the shallower depth necessary for ammonites), followed by a late rapid transgression immediately prior to, and during, the K-T transition. The timescale associated with the 30 foot gap and the duration of the late Maastrichtian transgression would have to agree to make this explanation plausible.

An increasing gravitational field, in conjunction with the above described rise in sea-level would compound the problem for ammonites. In other words, the effects would be magnified with increased gravitational change because the allowable depth limit (i.e., the implosion depth) for the ammonite (and nautilus) would be decreased. This would have the effect of limiting their habitats to the few remaining shallow areas during the transgressive phases and because surface gravity continued to increase beyond those phases, doomed the ammonites to eventual extinction.

Peter Ward and Phil Signor observed that ammonoids that survived extinction throughout Mesozoic crises had one thing in common; an especially thick siphuncle. From Figure 12-7 it can be seen that the siphuncle is the tube that runs along the inner surface of the outer shell through each chamber. As stated earlier, this tube is used to transport fluid/gases to the chambers to alter the buoyancy of the ammonite allowing it to maintain neutral buoyancy in the water column. According to Steven M. Stanley, in his book *"Extinction"*, in regard to the extinction/ siphuncle observation just mentioned, wrote:

"The functional reason for this pattern of extinction remains a mystery."

Since the siphuncle controls the ability to maintain neutral buoyancy in the water column, anything that would limit an ammonite's vertical mobility (e.g., a change in hydrostatic pressure with depth due to an increase in gravitational force) could result in the survival selection of ammonites with a more robust siphuncle.

122

Gravitational increases might cause an evolutionary pressure to strengthen the siphuncle as well as the shell itself. The shell structure of ammonites did change in the Cretaceous. They became more ribbed and developed tuberculation. Many paleontologists believe that these shell strengthening characteristics were a response to other shell crushing predators known as durophagous predators. While this is a reasonable assumption, the possibility of a response to gravitational change would also be a possibility. The gravitational influence on shell construction will be addressed later in this chapter.

Ward's 1983 *"The Extinction of the Ammonites"* (*Scientific American*) provides an extensive analysis of the shell design of ammonites. He describes the three shell types prevailing in the Cretaceous. They are the streamlined planispiral, ornamental planispiral and the heteromorphic shell type. Planispiral means a coiled shell within one plane. The heteromorphic shell consists of a variety of mostly non-symmetrical forms.

The heteromorph ammonites made up the majority of the ammonites at the end of the Cretaceous in terms of sheer numbers and species. And, they gradually deceased in size. If the ammonites were under increasing pressure from shell breaking fish and other predators that became more prevalent during the Cretaceous, it would be expected that the more streamlined, faster shell types would predominate. Instead, the smooth and streamlined shelled ammonites dwindled in numbers relative to the heteromorphs and the coarsely ornamented planispiral ammonites.

The answer to the question of why the nautilus was able to survive the K-T extinctions while the ammonite did not may be one of the vital clues to support the GTME as it applies to the K-T extinctions. Both creatures appear to be very similar in structure, behavior and habitat, yet only one survived. We also know that the late Mesozoic ammonites were dominated by heteromorphic species which had diverse odd shapes. If surface gravitation was gradually increasing we would expect structural changes which would aid the ammonites to descend to the same depth in the water column before the gravitational increase. Two things have to happen to accomplish this. First, the external shell has to be strengthened to maintain the same implosion depth that the ammonite had before the gravitational increase. Increasing shell thickness, developing ribs, tubercles and more robust spines would help. Late Mesozoic ammonites did develop many of these characteristics. However these changes alone would not be sufficient. The ammonite would have to make its buoyancy mechanism more efficient in order to counteract the increasing hydrostatic pressure associated with the gravitational increase. Based on that, the second change would be related to the way the siphuncle fills/empties the camerae with fluid to maintain the same level of neutral buoyancy.

Peter Ward, in his 1983 *Scientific American* article mentioned earlier, analyzed the structure of the ammonites relative to their buoyancy. He points out that the heteromorph ammonites, particularly the helically coiled ones, such as Turrilitidae and Nostoceratidae, had their siphuncles in a position so that they were in a constant relationship to the cameral fluid as their shells grew. This was in sharp contrast to the more standard planispiral forms where the relationship changed as the ammonite grew. He noted:

**"The heteromorphs seem to have gained
a more efficient   system for the maintenance
of neutral buoyancy at the expense of reduced
 streamlining."**

At the beginning of this chapter it was stated that some lineage of ammonites survived the K-T transition and persisted into the Danian Stage of the Paleogene Period. This is in conflict with the Alvarez Impact Theory of Extinction. Prior to the K-T, when ammonites faced extinction crises, the few survivors, finding themselves with little competition, were able to radiate and diversify with no trouble. Even after the Permian-Triassic extinction during which the recovery period lasted for a much longer period than that of the K-T, the ammonites were able to recover and flourish. So why would the ammonites that survived the Chicxulub impact continue on for tens of thousand, or hundreds of thousands of years before eventually going extinct? There must have been another factor, a much more powerful one, at work to eventually annihilate the ammonites. According to the theory described herein, that change was a gravitational one.

The proof of the survival of a lineage of ammonites past the Cretaceous can be found in a recent study (in 2005) of late Cretaceous ammonites in Europe. These ammonites are known as scaphitids. The study noted that at the Stevns Klint terminal Maastrichtian horizons, there were no signs of predation of the ammonoid fossil remains, thereby removing the predation cause of ammonite extinction. The areas studied were in Maastrichtian locations in central Europe in which 13 scaphitid taxa were found. One lineage in particular, Hoploscaphites constrictus, is focused on in the study because it prospered to the very end of the Cretaceous and one of its members, H.c. johnjagti subsp. nov., passed through the K-T boundary (also called the Cretaceous-Paleogene boundary). The researchers noted physical characteristics of the K-T surviving ammonites:

"dominated by individuals with a fully ribbed shaft and with ventrolateral tuberculation often extending to the aperture."

They note that ancestral H.c. populations have

"smooth flanks on the shaft and no tubercles in the adapertural region."

They conclude that the more heavily ribbed and tuberculated shells were a response to increased predation pressure, although this seems to contradict their findings at Stevns Klint.

THE GRAVITY THEORY OF MASS EXTINCTION

Further proof of the survival of ammonites beyond the bolide impact during the K-T transition is an article printed in the New York Times (11/6/07). Entitled *"Rethinking What Caused the Last Mass Extinction"*, by John Noble Wilford, the article describes the ammonites found above the global clay layer in the Manasquan River Basin in New Jersey, USA. The article points out that ammonites flourished for at least hundreds of years beyond the clay layer. Although most mass-media science writers attribute the clay layer to the bolide impact, some support its volcanic origin. Notwithstanding that controversy, Dr. Landman, invertebrate paleontologist at the American Museum of Natural History said:

"That's amazing and makes it hard to explain the ammonite abundances we find above the iridium anomaly."

The ammonite fossils shown in the article's photo consist of three scaphitid ammonites 1 ½" to 2" in diameter and a baculite specimen at least 6" long. The scaphitid ammonites display prominent tubercles very similar to the European hoploscaphites mentioned earlier. Why did the scaphitid species of ammonites outlast all the others? Within the context of the GTME, we can speculate about their longevity.

The scaphitid shell construction was more similar to the nautilus, generally planispiral having fewer whorls than that of the more complex heteromorphs. The shelled ribs became less pronounced to reduce weight and were replaced by tubercles to maintain shell strength. The shell started to unwind and twist away from its flattened disk-like form. The reduction in the number of whorls and unwinding of the shell were attempts to facilitate maintaining neutral buoyancy in line with Peter Ward's analysis of the decoupling of cameral fluid based on shell construction. However, the shell unwinding would weaken the overall shell's resistance to hydrostatic pressure, thereby reducing the implosion depth. Not having the nautilus's central siphuncle was most likely the

126

reason for the ammonites' unwinding strategy to deal with increasing gravitation.

As mentioned earlier, the present gravitational theory would explain the increase in shell strength characteristics, such as a fully ribbed shaft and extensive tuberculation as a gravitationally induced evolutionary change needed by ammonites to reach depths previously accessible before a rise in sea level in the late Albian Stage (~100 mya) of the Cretaceous. During this period, ammonites had to contend with both transgression and increasing gravitation; two factors limiting the availability of habitats at sustainable depths. Because this transgression was a relatively gradual one, ammonites were able to evolve various morphological forms, as the heteromorphs did, to compensate for the transgressive/gravitational pressure.

Gravitation increased at a higher rate toward the end of the Cretaceous. The surviving ammonites had to deal with a rapid transgression, part of a regressive/transgressive couplet, along with an even higher gravitation at the K-T transition; conditions which eventually led to their extinction. The seemingly counterintuitive idea of ammonite extinction during a period of transgression (at the Albian and K-T boundary) becomes more understandable if the gravitational factor is introduced.

The following chart shows the timescale of family members of the Ancyloceratida order of ammonites during the Cretaceous Period of the Mesozoic. The correlation between rising sea levels (i.e., transgression) and ammonite extinction is apparent. The two major periods of ammonite extinction occurred during the transgressions at the Albian(~100 mya) and very late Maastrichtian (~65.5 mya) Ages.

CRETACEOUS (ANCYLOCERATINA) AMMONITE EXTINCTIONS

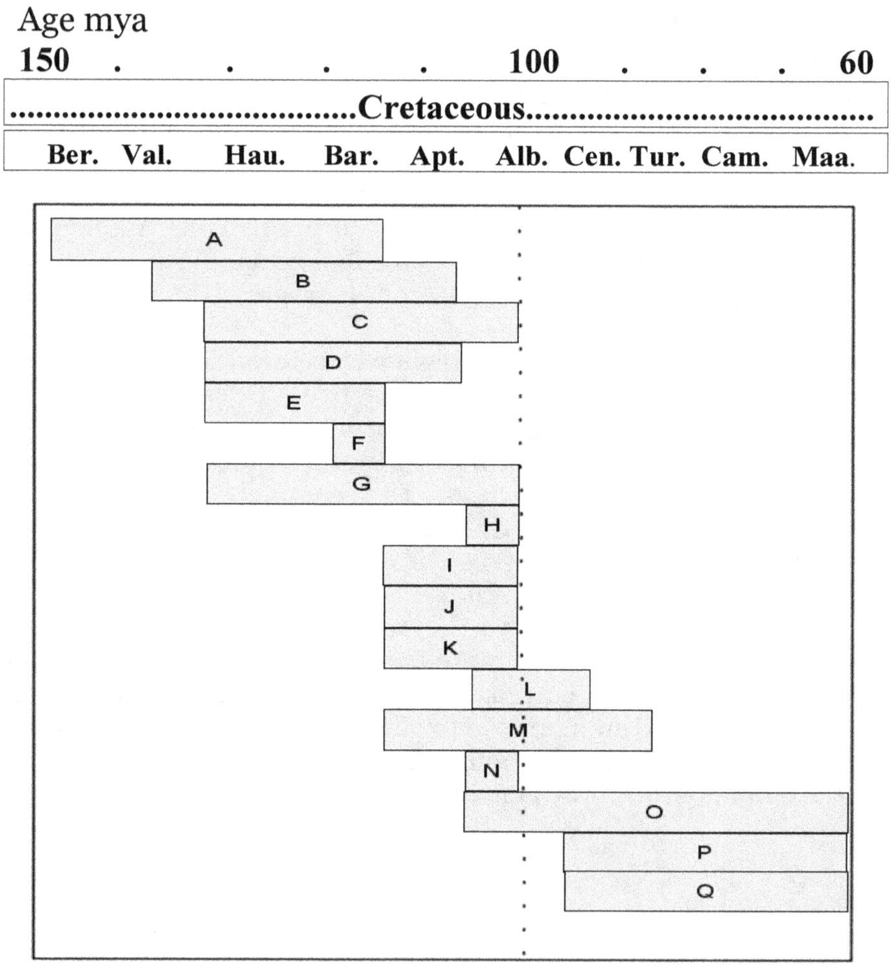
Age mya
150
100
60
Cretaceous
Ber. Val. Hau. Bar. Apt. Alb. Cen. Tur. Cam. Maa.
A
B
C
D
E
F
G
H
I
J
K
L
M
N
O
P
Q

FIG.12-9 Cretaceous ammonite extinctions

In FIG. 12-9 above, the timescale of the following ammonite families are shown:

A=Bochianitidae, B=Ancyloceratidae, C=Ptychoceratidae, D=Heteroceratidae, E=Hemihoplitidae, F=Hamulinidae, G=Douvilleiceratidae, H=Asteiriceratidae, I=Trochleiceratidae, J=Deshayesitidae, K=Parahoplitidae, L=Hamitidae, M=Anisoceratidae, N=Labeceratidae, O=Turrilitidae, P=Diplomoceratidae, Q=Nostoceratidae

THE GRAVITY THEORY OF MASS EXTINCTION

SUMMING IT UP

The ammonites almost went extinct during the P-T extinction (about 251 mya) when, as this author believes, release of methane from the ocean bottom poisoned 96 percent of the terrestrial and marine life. The recovery period took millions of years.

Another extinction crisis occurred at the Triassic-Jurassic boundary (about 200mya) when the ammonites were decimated again but were able to survive and prosper until a series of extinction events in the late Mesozoic. The author of the current theory believes that the T-J extinction events were caused by gravitational change initiated by the continental movements associated with the opening of the incipient Atlantic Ocean as described in *Chapter 13*.

Throughout the Mesozoic the ammonites evolved a wide diversity of shell configurations. The later ammonites, during the Cretaceous, developed extensive shell strengthening characteristics including a more heavily ribbed conch and the proliferation of tubercles. These characteristics became more pronounced as the K-T transition was approached and one logical explanation for these physical changes is a response to increased gravitation, which would have reduced the depth range of the ammonites.

The greatest rate of extinction of the ammonites during the Cretaceous, occurred at the Albian Age (i.e., the boundary between the upper and lower Cretaceous) and the K-T transition. Both periods were times of significant transgression. A rapid trangression combined with a rapid increase in gravity would have a combined effect to reduce the locations at which ammonites could proliferate. Peter Ward's conclusion about the missing ammonite fossils in the 30 foot gap at Zumaya being due to excessive depth, makes the transgression/gravity-increase scenario a reasonable explanation for the final demise of the ammonites.

One lineage of ammonites, Hoploscaphites constrictus johnjagti subsp. nov., passed into the Danian Age of the Tertiary. Although the faunal recovery period beyond K-T was relatively rapid compared to that at the end of the Permian-Triassic extinction, the K-T surviving ammonites eventually died out. Why was this time different? Why didn't they prosper again, having survived the K-T boundary? The same question applies to the ammonites found above the clay layer in the Manasquan River Basin in New Jersey, USA.

The chart on the following page displays the Cretaceous ammonite extinctions in relation to the salient parameters based on the GTME, namely an increasing gravitational field at the Earth's surface and major sea-level transgression. Data for ammonite abundance is derived from Peter Ward's "*The Extinction of the Ammonites*", *Scientific American*, 10/83.

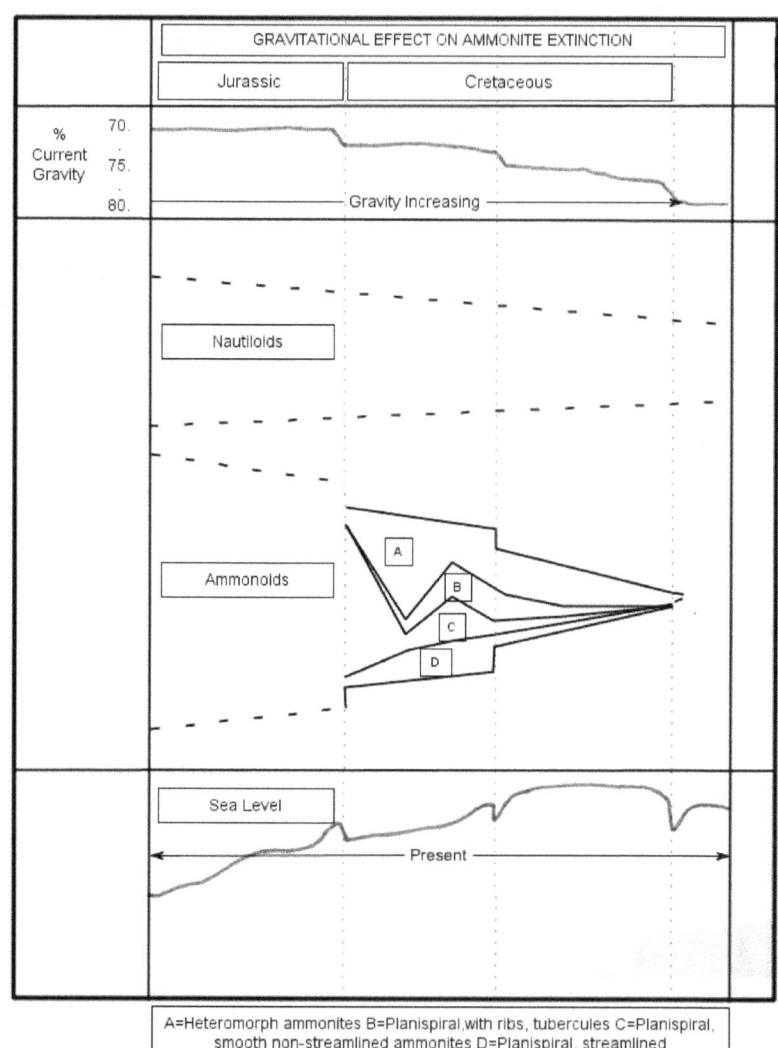

A=Heteromorph ammonites B=Planispiral,with ribs, tubercules C=Planispiral, smooth non-streamlined ammonites D=Planispiral, streamlined

FIG. 12-10 Ammonite extinction during late Mesozoic relative to gravitation and sea-level.

As can be seen in the above representation of extinction, the nautiloids experienced a milder extinction rate compared to the ammonoids. These extinctions were much more severe during periods of sea-level transgression because the combination of increasing gravitation and increasing sea-level had a cumulative negative impact in reducing available shallower habitats for both but was especially detrimental for the ammonites. The scale for the % of current gravity is, of course, an estimate but is probably not far off.

The nautiloids survived the K-T extinction period and exist today. The nautiloids have a thicker shell than did the ammonoids, have fewer shell whorls than most of the ammonites and perhaps most important have a centrally located siphuncle which, based on the analysis above, would facilitate the transfer of cameral fluid facilitating neutral buoyancy. Therefore, although the nautilus was negatively affected by the increasing gravitation, it was better positioned to withstand the change. Peter Ward commented:

> **"The low diversity of modern nautiloids suggest however, that the time for chambered shells, no matter how well engineered is past."**

What is different in today's environment in comparison to the distant past that would make this true? Shell-less cephalopods such as the squid and octopus, which occupied the same environment as the ammonites, did not become extinct.

The GTME posits an increasing gravitational field at the surface of the Earth during the Mesozoic including pulses (i.e., rapid accelerations), that were more profound as the Mesozoic drew to a close. It is believed the extinction of shelled cephalopods including the ammonite was the direct result of the those gravitational increases enhanced by sea-level transgression. The extensive diversity of ammonite shell types was an attempt to compensate for increasing gravitation. In addition, the new heteromorph morphology, which aided vertical mobility became a handicap for

---

133

migrating to distant habitats. The reason for the survival of the nautilus and not the ammonite is probably the location of their siphuncle coupled with a stronger shell. The central siphuncle location in the nautilus was better positioned to more efficiently fill/empty its camerae when gravitation increased, thereby facilitating neutral buoyancy. The accelerating gravitational field strength would have been caused by the rapid separation (especially east-west movement) of the continents.

FORAMINIFERA.................................

Foraminifera (Forams for short), are unicellular organisms that are known from the earliest Cambrian of about 550 mya. They are shelled marine organisms and the shells are referred to as "tests." Forams have been described as "an amoeba with a shell." See photos on page 67. The tests of the earliest forams were soft until the Devonian (400 mya) when the (calcareous foraminiferan) fusulinids flourished in the latter Carboniferous and Permian Periods (of about 300-250 mya). The fusulinids became extinct at the end of the Permian. The textularinids and the rotalinids, most likely the descendants of the Carboniferous miliolids, radiated throughout the Mesozoic Era.

Forams are divided into two groups, the benthic (bottom dwelling) and the planktic (surface or near-surface) forms. The earliest planktic forams were small, rounded types. The tests of the planktic forams are made of thin sections of calcium carbonate which trap air providing them with buoyancy. As they diversified, they developed a wide variety of complex forms including the "popcorn", triangular and ridge shapes. The planktic forams underwent a massive extinction during the K-T transition at the end of the Mesozoic. The forams that survived were similar to the early planktic forms; they were small and rounded. There was some recovery in the Paleogene Period including the return of elaborate shaped planktic forams. During the Eocene (about 38 mya) and the Miocene (about 23 mya) these more complex forams were decimated again with the small rounded shapes surviving. Those were times when polar ice caps were forming and subsiding. Today, the planktic forams are found primarily in lower latitudes.

Large lentil-shaped forams have been found in the granite blocks of the pyramids. Due to their widespread marine habitats and their relatively rapid turnover and diversification, forams are used extensively in biostratigraphy and paleoclimatological research. The ocean temperatures for a period can be estimated by the oxygen isotope ratios obtained from the foram microfossils. Recently, the planktic foram fossils have been studied to obtain a more precise timing for events near the K-T transition. This process obtains what is called biozonation.

The change in the size and structure of the planktic forams across the K-T boundary is well documented. They went from large fancy forms to small simple, mostly rounded forms. What is the reason for this change? One explanation is based on the premise that marine life-forms were under extreme environmental stress at that time. This would have resulted in the 'Lilliput Effect' thereby favoring smaller members of a species.

The GTME offers a different explanation for the foram change. Since the planktic forams have shells, or tests as they are called, that trap air to maintain buoyancy, those forams that had excess shell material which is made of calcium carbonate, like the highly ornamental ones, were at a disadvantage with an increasing gravitation and therefore more likely to become extinct. Since the forams did not require extensive vertical mobility in the water column as did the ammonites, their extinction was delayed to the very end of the Cretaceous. Also, as noted elsewhere in this book, the K-T foram extinctions were most heavy in the low latitude locations, areas which would have had the greatest rate of increase in gravitation.

CHAPTER 13: THE TRIASSIC- JURASSIC EXTINCTIONS

Also known as the end-Triassic, this extinction occurred at the Triassic-Jurassic boundary (about 200 mya) and is purported to have wiped out 20% of all marine families, and on land, most non-dinosaurian archosaurs, most therapsids, and the last of the large amphibians. The final extinction of the conodonts occurred at that time. Apparently, the terrestrial extinctions cleared the way for the rise of the dinosaurs. The end-Triassic mass extinction is one of the five most catastrophic in the Phanerozoic.

Several possible explanations for the extinctions have been given. Sea-level changes, bolide impact, flood basalt volcanism and climate change are often cited. No known temporal impact site rules out the bolide impact explanation which several theorists proposed because of the suddenness of the event. The Manicouagan Crater, thought to have been formed by bolide impact at the T-J boundary has been dated much earlier based on sedimentary evidence. A duration of ten thousand to fifty thousand years has been estimated for the extinction period although there is strong evidence for a series of extinction pulses preceding the T-J boundary. The volcanic flood basalts of CAMP (Central Atlantic Magmatic Province) were active near the T-J boundary but whether they were synchronous with the extinctions is still being debated. The pro-volcanism supporters claim that the release of CO_2 (causing warming) and SO_2 (causing cooling) would have stressed the environment enough to cause extinctions.

However, Lawrence H. Tanner, of Bloomsburg Univ.,Pa, USA, recently studied the isotopic composition of contemporaneous fossil soils and found no evidence for altered CO_2 levels of the atmosphere. Tanner said:

"Other possibilities need to be investigated more fully."

THE GRAVITY THEORY OF MASS EXTINCTION

Examination has been made of the CAMP lava flows in Europe and North America (by Fowell and Olsen, 1993) which overlay sedimentary formations that contain earliest Jurassic fossils and pollen. The conclusion drawn is that the entire CAMP volcanic activity postdates the extinctions by thousands of years.

Sea-level changes have often been invoked to explain many extinction periods, including this one. They do not explain the synchronous terrestrial extinctions, therefore those theories are based on a weak foundation.

During the T-J transition there was a strong regressive/transgressive couplet. This rapid drop in sea-level followed by rapid rise appears to be common during extinction intervals. Except for periods during the formation of ice caps and their subsequent melting, scientists don't offer an explanation for the rapid couplets. The GTME offers an explanation for the regressive/transgressive couplets. *Chapter 15 Sea-Level Changes And Surface Gravity* will describe the current theory's explanation for this phenomenon.

Hallam & Wignall (1997) point out that in certain sections there were significant T-J plant extinctions. Mentioned specifically are seed ferns Glossopteridaceae, Peltaspermaceae and Crystospermaceae. They note significant palynofloral turnover in Canada, Europe and north-eastern United States. Also emphasized is that there was no corresponding plant extinctions in northern Siberia or Australia, two regions where the palynomorph record is good. Can one conclude that there was a latitudinal bias toward T-J plant extinctions? If so, this is additional support for the GTME.

The T-J extinctions are explainable by the GTME. Major rifting of Pangea began in the late Triassic. The creation of the Atlantic Ocean was beginning with the east-west separation of the continents. As explained in this book, that rapid separation of the continental tectonic plates would induce a movement of the Earth's cores toward a "normal" central position resulting in two things. First, a gravitational increase pulse resulting in selective extinctions. The extinction of the large amphibians would be an example. The

---

138

therapsids went extinct at this time and paleontologist Peter Ward, not with gravity in mind, made an observation that indirectly supports it. He said:

"Perhaps creatures reproducing with buried eggs survived and large animals with live births did not."

Second, that rapid core shift would induce a pulse of core/mantle volcanism (i.e., a plume) which would lag the extinctions due to the time it would take for the plume to rise from the core/mantle boundary. If the CAMP did in fact postdate the extinctions, then the GTME would be a viable explanation for the T-J extinctions.

PART V: DENOUEMENT

CHAPTER 14: THE EARTH'S CORE AND SURFACE GRAVITY

The basis of the GTME (Gravity Theory of Mass Extinction) can be visualized as shown in Fig.14-1.

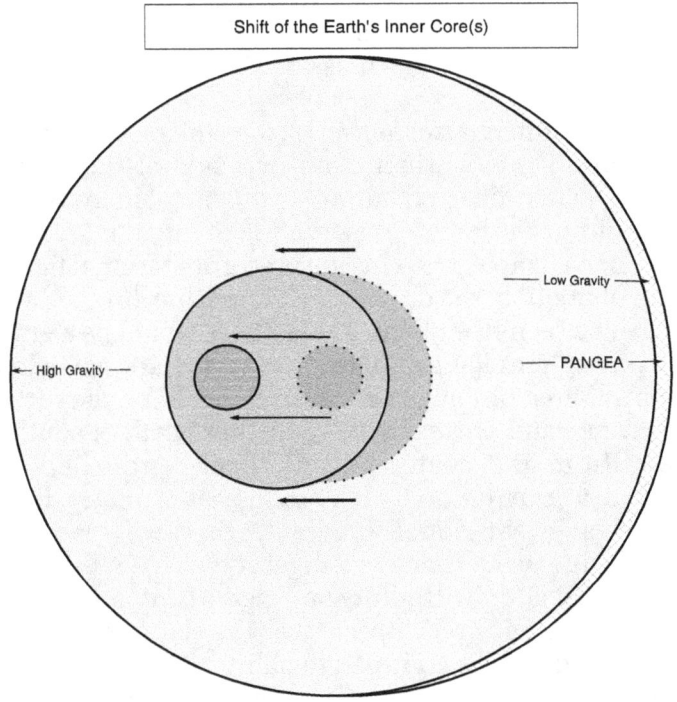

FIG. 14-1 Gravitational effects from core shift

DENOUEMENT

As the continents coalesced during the Permian Period, the Earth's inner/outer iron cores shifted from a central position away from Pangea. This core shift increased the distance between the core and the Pangean surface and therefore, based on Newton's Law of Gravitation, lowered the force of gravity at the surface:

$$g = KMm/d^2$$

g is force of gravity at surface on a mass m (i.e., weight of m)

K is a constant (usually called G)

M is mass of Earth

d is distance between the center of mass of M and m

The core shift is a response to the unbalancing effect of the consolidation of the continents on one side of the globe to form Pangea. The scientific explanation of this movement is a subject of physics including the law of Conservation of Angular Momentum and will not be delved into here. This lowered gravitational field led to the gigantism of not only the dinosaurs but marine animals and flora as well. As can be seen from Figure 14-1, there would be a gravitational gradient throughout Pangea. The lowest gravitational field would be at the central area of Pangea because it would be the furthest from the shifted core and areas that were further north or south, and east and west, from the center of the super-continent would have experienced a gravitational field that was stronger than central Pangea. This gravitational gradient, that was more apparent latitudinally due to the shape of Pangea, helps to explain many of the anomalies associated with dinosaur gigantism and extinction. For example, as noted in *Chapter 10*, the largest dinosaurs, the sauropods, tended to be confined to habitats closer to the equatorial region of Pangea even though other dinosaurs inhabited all continents including Antarctica. Also, as described in another part of this book, various K-T extinctions, such as the planktic foraminifera, were most severe in the equatorial regions. This occurred because when Pangea broke up, the cores began to return to a more central location, thereby increasing the gravitational pull at the surface of Pangea. It can be seen from Figure 14-1 that the greatest effect of this increasing gravity would be felt in the equatorial region.

During the Mesozoic Era, a period of some 185 my, Pangea gradually broke up. Each movement of a continental tectonic plate had an effect on the movement of the inner/outer cores, primarily to return them to it's a more central location. Therefore, each movement caused a pulse of gravitational change. Those pulses were primarily increases in gravitation due to the separating of the plates. The rapid movements corresponded to pulses of gravitational change. These pulses caused corresponding extinction pulses which correspond to end-Triassic, end-Jurassic and the stepped extinctions which occurred during the Cretaceous Period.

The basis of the GTME is the position and movement of the continental tectonic plates. Many references to pulses of increasing gravitational fields at the Earth's surface along with the concomitant extinction periods have been posited. Therefore, those major plate movements must be identified. That will not be an easy task because there is an amount of uncertainty about exact timing and movement of the continental plates.

An example of the uncertainty in establishing the position of the Pangean sub-continents can be found in an article "*Study Solves Pangea Puzzle*" (*SCIENCEDAILY* 12/19/00). University of Michigan and Geological Survey of Norway researchers attempted to reconcile the different models of how the continental components of Pangea fit together. One model, known as Pangea "A" and is the one most frequently depicted shows the South American continent directly south and in contact with North America. The Pangea "A" model seems to be corroborated by fossil and mountain range data. However, when geologists construct a model (Pangea "B") based on paleomagnetic data, they arrive at a model that indicates the South American continent was much further east with its northwest coast in contact with the east coast of North America. The researchers conclude that the Earth's magnetic field is not "perfectly dipolar" at all times and the paleomagnetic model for other times and continental configurations must be reevaluated.

DENOUEMENT

The interesting point made in the research study about the Earth's dipole is that the GTME predicts a shifting dipole. A shift in the Earth's cores as described in this book would, by definition, shift the axis of the magnetic dipole away from its current, Earth-centric location. And, that shift would have been greatest when the Pangean super-continent was fully consolidated.

Keeping in mind the above caveat, there are some generally recognized movement/position data of the continental tectonic plates. Below is the general timeline.

PERIOD (mya)	CONTINENTAL MOVEMENT
Late Carboniferous 300	Formation of Pangea was well underway. The North American/European land masses had collided with Gondwana (i.e, southern continents). There would have been reduced surface gravity and this seems to be confirmed by the appearance of 5 foot long millipedes (e.g., Arthropleura) and dragonflies (e.g., Meyaneuropsis permiana) with two and one half foot wingspans.
Perm-Triassic 251	Pangea at or near total consolidation. **Mass Extinction** about this time. Author believes low gravitational environment released methane from submarine methyl hydrates.
Triassic-Jurassic 200	Pangea begins breakup. Laurasia/Gondwana begin rifting. Atlantic Ocean starts expanding. T-J **Mass Extinction.**
Jur.-Cretaceous 145	South America splits from Africa. Tethys Sea begins to close. Laurasia moves clockwise. Africa moves north with counterclockwise rotation. Period of **Extinction**...notably sauropods.
Cretaceous 145-65	South America and Africa continue east-west separation, Atlantic Ocean expands, India moves east then north. **K-T Mass Extinction.**

DENOUEMENT

THE CIMMERIAN PLATE........THE KEY THAT UNLOCKS THE DOOR TO THE MYSTERY OF PANGEA'S BREAKUP

The Cimmerian Plate, a relatively obscure continental tectonic plate may provide the answer to important questions concerning not only the initial cause for the breakup of Pangea but also gravitational effects which support the GTME.

The Cimmerian Plate was formed approximately 300 mya when the southeastern edge of Pangea began to rift and separate from the super-continent. This long plate, also called a micro-continent, began to sweep across the Paleotethys Ocean with its northwestern edge acting as a pivot point and the rest of the plate rotating counterclockwise.

The GTME supports the theory that the P-T extinctions (~251 mya) were the result of the disassociation of methane from the submarine methyl hydrates (see *Chapter 11*). It can be seen from Figure 14-2 that as the Cimmerian Plate rotated to a position parallel to, and near, the paleoequator, the gravitational effect, as described in the GTME, would be to lower the Pangean surface gravity even further. Another important point which was not mentioned earlier is the northward migration of Pangea at this time. Therefore, the triple effect of the consolidation of Pangea, its northward movement to a more central equatorial position and the pendulum-like swing of Cimmeria across the paleoequator would have resulted in the lowest surface gravity of the Paleozoic Era. This lowering of surface gravity, based on the GTME, is the trigger that released the methane.

FIG. 14-2 Location of Cimmerian Plate at the end Of the Carboniferous Period

DENOUEMENT

Once the Cimmerian Plate had rotated north of the paleoequator, the effect would be reversed (i.e., a gravitational increase). The Cimmerian Plate would then collide with the south China Blocks, or microcontinents as they are sometimes called, moving itself as well as those microcontinents to an eventual collision with Laurasia. This collision is estimated to have occurred about 200 mya, the period of the Triassic-Jurassic extinctions. This is consistent with the GTME, which posits a rapid shift in the Earth's cores at that time to a more Earth-centric position. As explained in this book, that rapid core movement would initiate, in the following order, these events:

- A synchronous extinction pulse, more severe for terrestrial biota than marine biota.
- A synchronous initiation of a large mantle plume which would not reach the Earth's surface for an extended period of time of perhaps hundreds of thousands to a million years.
- A synchronous regressive half of a regressive/transgressive couplet that would also occur over an extended period of time. The transgressive half would be the cause of a second extinction pulse for marine organism that depended on vertical movement in the water column (e.g., ammonites).

Finally, the CAMP flood basalt eruptions, initiated by the Cimmerian Plate movement as described above would have initiated the creation of the Atlantic Ocean along with the breakup of Pangea.

148

THE END DEVONIAN (FRASNIAN-FAMENNIAN) EXTINCTIONS

As mentioned earlier, most of the attention in this book is focused on the three youngest mass extinction events. Since the GTME is a unified theory of mass extinction, the End-Devonian mass extinction must be examined, at least briefly.

Although there was some controversy about the duration of these extinctions, it is now accepted that they took place over a long interval of perhaps 20 my and might have been a series of extinction pulses. Also recognized are two major extinction peaks referred to as the Kellwasser (378 mya) and Hangenberg (360 mya) Events. The most common explanation for these extinctions has been proffered but no single one is strongly supported primarily because of the long duration of the extinction period. Bolide impact has been rejected outright not only because of the duration issue but also because no evidence, such as iridium or widely distributed tektites, have been found.

Glaciation is sometimes cited as the extinction mechanism because the southern Gondwana super-continent had moved above the southern pole during the late Devonian. However, the timing of the glaciation seems to be nearer the Devonian-Carboniferous boundary (Hallam & Wignall, 1997). During the Devonian, the arrangement of the continental landmasses was different from that of the Mesozoic. The Euro-America continent was positioned near the equator and the much larger Gondwana super-continent was positioned in a high southern latitude near, and sometimes above, the southern pole. It gradually moved north in the early Devonian well away from the southern pole. Later in the Devonian, the exact timing doesn't seem clear, Gondwana reversed course and moved south so that the South American expanse was situated above the southern pole. Based on one representation (Steven M. Stanley, 1987), this polar overlay took place starting about two thirds through the Devonian. This would be near the time of the Kellwasser Event.

The movements of Gondwana, as explained by the GTME, would cause significant changes to surface gravitation. When Gondwana moved over the southern pole, surface gravity would increase and the

149

opposite would occur when it reversed direction and moved north. As with most of the other mass extinctions described in this book, the following characteristics of the late Devonian extinctions would be predicted by the GTME:

1. Low latitude life forms would be negatively affected compared to those at higher latitudes due to the latitudinal gravitational gradient.
2. Rapid eustatic fluctuations in sea levels (i.e., regressive/transgressive couplets) would occur dependent on the rate and direction of continental movement.
3. Marine life that depended on vertical mobility in the water column would be most negatively affected by increased gravitation first, followed by marine life that floated near the surface. Benthic (i.e., bottom dwelling) marine life would be least affected (when surface gravity increased).
4. When Gondwana moved north away from the southern pole, surface gravitation decreased. This would have been more detrimental to benthic life; just the opposite of above.

All of the above are observed symptoms of the late Devonian extinctions.

Finally, the late Ordovician extinctions are briefly described in *Chapter 1*. What is significant is the comparable movement of Gondwana over the southern pole and the rapid regressive-transgressive couplet at the time of those extinctions. All of these factors support the GTME and therefore, the unification of the gravitational causal mechanism for all five mass extinctions.

CHAPTER 15: SEA-LEVEL CHANGES AND
SURFACE GRAVITY

The GTME posits that a reduced surface gravity existed on Pangea. Based on the shifting core scenario, the antipodal region of the Earth would have had a corresponding increase in surface gravity. Since sea levels would be dependent on the strength of surface gravity, a higher sea-level near Pangea and a lower sea-level opposite the super-continent would have occurred. As with the latitudinal and longitudinal gravity gradient, there would have been a corresponding sea-level gradient. The following drawing depicts this.

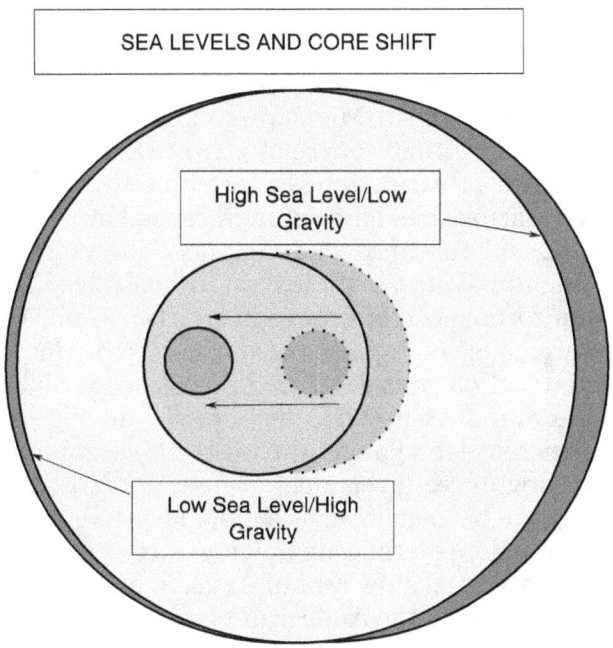

FIG. 15-1

DENOUEMENT

During the Mesozoic and Paleozoic Eras there were many instances of rapid sea-level drop/rise sequences. These are called regressive/transgressive couplets and many of these were eustatic (i.e., worldwide). Those that have studied this phenomenon know that the only logical explanation for the rapid regressive/transgressive couplets is the formation and melting of polar ice caps or large ice sheets. They also know that the Mesozoic was a period of unusually elevated temperatures and was devoid of large ice formations. What can be a reasonable alternate explanation for couplets? Science professionals have offered several possible explanations. The formation and melting of Mesozoic ice sheets, Milankovich cycles and TPW (True Polar Wander) have been suggested. None of the above explanations can explain the coincidence of terrestrial and marine extinctions associated with the regressive/transgressive couplets.

Shanan Peters, assistant professor of geology and geophysics at the University of Wisconsin-Madison, in 2008, noted that changes in ocean environments that were related to sea level exerted a driving influence on rates of extinction. He attributed the extinctions to changes in the marine shelf environment caused by sea level change.

The GTME explains this phenomenon as follows. The movement of the continental tectonic plates do not always proceed at a continuous uniform speed but move in spurts as well. Those spurts or accelerations cause high pulses of surface gravity increases when plates move latitudinally as described in this book. Since there was a global differential sea-level, as shown in Fig. 15-1, those gravitational increases on Pangea during the Mesozoic and Paleozoic caused a lowering of sea levels near Pangea and a rise in sea levels antipodally. It can be visualized as a global wave forming slowly, in geological timescale, receding near Pangea (i.e., regression), then followed by the eventual slow rebound (i.e., transgression).

Norman Newell, of the American Museum of Natural History, New York, observed the coincidence of periods of extinction and sea-level regression. His proposal for lowered sea levels as a cause for extinction was based on the fact that most of the marine biota lived in the shallow epicontinental seas and any significant lowering of the sea-level would reduce their habitat. Newell specified the following periods of coincidences of extinction/sea-level regression:

1. End Ordovician

2. Late Devonian

3. End Permian

4. End Triassic

5. End Cretaceous

A. Hallam and P.B. Wignall are well known experts in the subject of Phanerozoic sea-level change. In their 1999 "Mass Extinctions and Sea-level Changes" publication they noted:

> **"Rapid high amplitude regressive-transgressive couplets are the most frequently observed eustatic changes at times of mass extinction, with the majority of extinctions occurring during the transgressive pulse....."**

The above observation by Hallam and Wignall was used to support the belief that anoxic bottom water might have played a part in the extinctions. The above observation supports the GTME as described in this book. The regressive half of the couplet would indicate a rapid increase in surface gravity accompanied by immediate terrestrial extinctions and to a lesser extent marine extinctions. The followup transgressive half of the couplet would, as in the case of the ammonites, eliminate habitats with the aid of increasing gravity by reducing their vertical mobility in the water column because higher surface gravity would increase the density of water.

The author of the GTME believes that the Permian-Triassic extinction was not caused by sea-level change (see *Chapter 11*) even though there was an extreme drop in sea level near the P/T boundary. The GTME explains this massive regression as a rapid return of the Earth's core elements to their current geocentric position resulting in a major pulse of increasing surface gravity. As noted elsewhere in this book, the extreme drop in sea level and the warm temperature at that time released massive amount of methane

DENOUEMENT

from the sea bottom. Therefore there was a dual causality for the P/T extinctions, increasing surface gravity and methane release. It is doubtful whether there were any Mesozoic extinctions due exclusively to sea-level change when there were both synchronous terrestrial and marine extinctions.

During the Mesozoic Era sea levels were very high relative to today. The following is a graph of sea levels (from the Hallam et al., Curve and the Exxon Sea Level Curve).

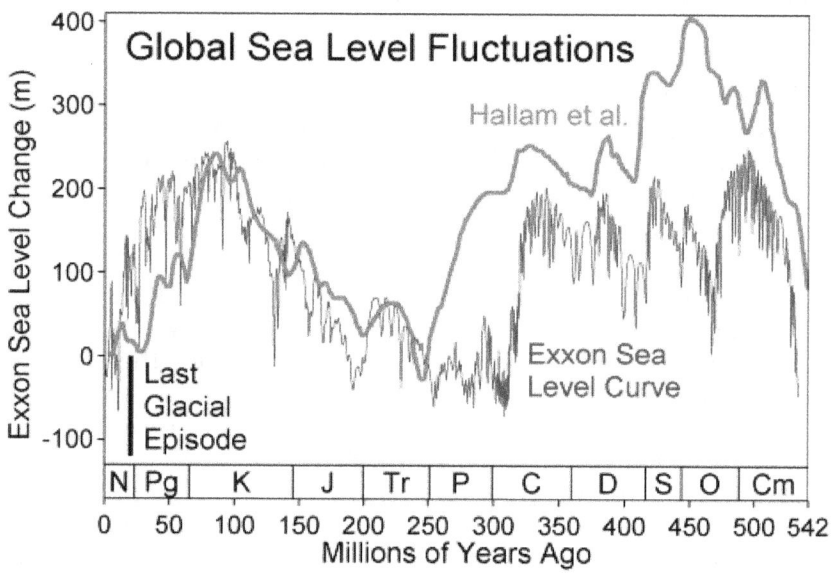

FIG. 15-2 Sea-Level Curves Hallam et al.
and Exxon Sea Level Curves

The Mesozoic regressive/transgressive couplets that are shown on the curve correspond with extinction periods. The linkage to gravitational change, as theorized by the GTME, is apparent.

There are several factors that contributed to the high sea levels in the Mesozoic:

1. The formation of volcanic ridges and plateaus on the bottom of the ocean.

2. The periods of increased sea floor spreading which caused the thermal expansion of the lithosphere adjacent to the spreading centers.

3. The thermal expansion of the oceans' water. Warmer water occupies more space than cooler water.

4. Lower surface gravity (per the GTME). It would also entail higher antipodal sea levels.

When viewing the sea-level curve above, the levels displayed in the graph are a summation of all four components. The current theory posits that each **rapid drop** in sea-level, because there were no known verifiable ice formations during the Mesozoic, were the results of latitudinal spurts (i.e., accelerations) of continental tectonic plate movements.

> **The GTME posits that most, if not all, regressive/transgressive couplets during the Mesozoic and Paleozoic were caused by pulses of increasing surface gravitation on Pangea due to latitudinal movement of continental tectonic plates. All major regressive/transgressive couplets were periods of major extinction, both terrestrial and marine. Therefore, those extinctions were not primarily caused by sea-level changes but by the effects of increased gravitation. However, the transgressive phase did cause selective marine extinctions.**

CHAPTER 16 SUMMATION

The Gravity Theory of Mass Extinction (**GTME**) can be summarized as follows:

1. The coalescing of the continents during the Permian Period was a gradual process that lowered the gravitational field on the surface of Pangea due to the linkage between tectonic plate movement and the shifting of the Earth's inner/outer iron cores. That linkage is governed by scientific principles, one of which is the Conservation of Angular Momentum.

2. The northward movement of Pangea caused surface "g" to increase to near current value causing a rapid drop in sea level. Combined with warm temperatures caused a release of methane from the methyl hydrates at the bottom of the sea toward the end of the Permian Period which led to the extinctions before and through the P-T boundary (of about 251 mya).

3. The lowered gravitational field during the Mesozoic Era allowed the terrestrial animals (especially the sauropods) to reach their gigantic size after their competitors were eliminated in the Triassic-Jurassic extinctions. Marine animals and flora also experienced gigantism.

4. As Pangea split up, which it did over a long period of time, there were corresponding gravitational increases at the surface of Pangea. Rapid plate movements (i.e., accelerations) caused pulses of increased gravitation and corresponding pulses of extinction. Those pulses correspond to the end-Triassic, end-Jurassic and end-Cretaceous extinctions. The strength of these pulses were primarily determined by the size and direction of the continental tectonic plate movements and the felt effects were dependent primarily on the latitudinal position on

Pangea; the strongest were in the equatorial region. This linkage between tectonic plate movement and the shift in the Earth's inner/outer cores is described in this book.

5. Each gravitational increase had a corresponding effect on the terrestrial and marine life-forms. Most notable is the decline in size of the sauropods after the late Jurassic Period.

6. The breakup of Pangea was especially rapid toward the end of the Mesozoic Era and the gravitational-increase pulses were instrumental in causing the rapid stepped extinctions noted in the Cretaceous. Among these extinctions are those of the inoceramids and ammonites.

7. Marine animals that had shells became extinct in a sequence that was directly related to the size of their shell relative to their total weight and to their mobility in the water column. The large clam-like inoceramids went extinct by the mid-Cretaceous, belemnites and ammonites gradually became extinct from the mid-Cretaceous Period through the early Danian of the Cenozoic Era. Microscopic marine organisms such as planktic foraminifera, which floated near the surface and therefore did not require the effort to vertically ascend/descend the water column, suffered extinctions later toward the end of the K-T transition. Those forams that went extinct were those most susceptible to a gravitational increase; the large, very ornate, many-chambered variety. In other words, those organisms with excessive calcareous (i.e., calcium carbonate) structure were more vulnerable because they were less buoyant as a result of the increasing gravitational field.

8. In the terrestrial realm, dinosaurs faced a dual obstacle; an increasing gravitation and mammal predation. Those that had a physical (both in size and internal)

structure that could not quickly evolve to a higher gravity tolerant form went extinct. Those extinctions allowed an increase in predatory mammals which were, contrary to what many believe, a direct threat to egg-laying dinosaurs. It has often been written that the mammals of the Mesozoic were small shrew-like creatures but the discovery of the carnivorous dog-size Repenomamus fossils in China falsify this claim. It was the diversity of dinosaurs, especially the smaller carnivorous ones, during most of the Mesozoic that limited the growth of mammals. Unlike present day crocodilians, that bury their eggs relatively deeply to protect them from predation, dinosaurs appear to not have done so. Birds, the avian dinosaurs, were able to hide their eggs just as they do today and are still here. Therefore, even the smaller non-avian egg-laying dinosaurs that one would expect to survive an increasing gravity environment went extinct, including the small burrowing dinosaurs whose fossils were recently discovered.

9. In both the terrestrial and marine realm, animals that carried their young for extended periods of time were most susceptible to gravitational extinction. In the northern hemisphere, most if not all, marsupials became extinct. In the seas, the reptile ichthyosaurs which did not lay eggs put gave live birth to young went extinct before the mid-Cretaceous. The mosasaurs, slower swimmers, also gave live birth and went extinct in the late Cretaceous.

10. The massive flood basalt volcanism at the Deccan Traps, Siberian Traps, the NATIP (North Atlantic Tertiary Igneous Province) as well as all of the other major flood basalt episodes were initiated by the shifting of the Earth's core in response to the movement of tectonic plates. This explains why massive flood basalt volcanism has tapered off since the Mesozoic as the inner core has returned to its normal central position. This also explains why many extinction periods are temporally linked to massive flood

basalt volcanism. Massive flood basalt volcanism is an after-effect of the, especially rapid, core movements and gravitationally induced extinction; not the cause of the extinction. This is why extinction periods began just before the Triassic/Jurassic CAMP and the Cretaceous/Cenozoic Deccan eruptions.

11. The bolide impact near the K-T boundary may have been initiated by the wobble of the Earth due to the shift in the Earth's plates. When the cores moved away from a central position during the formation of Pangea throughout the Carboniferous, it did so at a very slow rate and therefore any wobbling at that time was minimal. When the cores started to move back to a central position during the Mesozoic, the movement was relatively rapid during the late Mesozoic due to the relatively rapid breakup and dispersal of the continental tectonic plates. This created maximum wobble of the Earth. This wobble could have dislodged the asteroids which were orbiting in a near Earth orbit and were probably originally part of the 298 Baptistina asteroid estimated to have broken up 160 mya. Scientists believe that another asteroid from this same parent asteroid collided with the Earth's moon about 109 mya forming the crater Tycho. There may have been other impacts near the K-T boundary beside the Chicxulub event due to this wobble.

12. The lowered gravitational field on the surface of Pangea during the Mesozoic Era would, by definition, have caused a corresponding increase in gravitational pull on the opposite side of the globe where there were no land masses. The sea-level around the globe would have been affected. There would have been higher sea levels near Pangea and the effects would have varied latitudinally and longitudinally because of the gravitational gradient. In general, lower sea levels occurred antipodally to Pangea and higher sea levels occurred near Pangea. Other

160

powerful factors also came into play, such as submarine volcanic ridge formation and active oceanic ridge spreading, to form the high sea levels observed during the Mesozoic.

Science professionals have pondered the late Maastrichtian regression. It happened so rapidly that an ice cap formation scenario had been proposed even though elevated temperatures at that time make that unlikely.

The massive and rapid late Maastrichtian regression can be explained by the reduction in the differential gravitation between both sides of the globe during the rapid breakup of Pangea at the end of the Cretaceous. A gravitational increase pulse would cause a rapid reduction in sea-level near Pangea and result in a eustatic sea-level approaching a more "normal" worldwide balanced level.

Pulses of increasing gravity would also explain the many regressive/transgressive couplets which occurred during the Mesozoic. The regressive part of those couplets were synchronous with both terrestrial and marine extinctions and the transgressive part with marine extinctions due to the reduction in the depth range caused by the combined gravitation/transgression effect.

13. Based on the points made above, all of the major reasons given for the K-T extinctions, namely bolide impact, massive flood basalt volcanism and massive sea-level regression can be directly attributed to an increasing gravitational field on Pangea. In other words, what seems like a convergence of coincidences are in actuality, expected side effects consistent with the Gravity Theory of Mass Extinction.

DENOUEMENT

14. There may be a relationship between the movement of the Earth's inner core as described in the current theory proposed and polar magnetic reversals. On average, there is a pole reversal about every 500,000 years. Because there was an unusually long period in the Mesozoic Era when there were no reversals of the poles (i.e., a long chron) which also coincided with the rapid shifting of the core, as proposed by the GTME, there may be an influence on polar magnetic reversal by the movement of the inner core and, by extension, movement of tectonic plates.

15. The movement of the Cimmerian Plate across the Paleotethys Ocean was instrumental in lowering the Earth's surface gravity based on the inner-outer core/tectonic plate basis of the GTME. The resultant lowering of gravity was partially responsible for the release of the methane from the submarine methyl hydrates initiating the Permian/Triassic mass extinction.

16. The above movement of the Cimmerian Plate as it rotated above the paleoequator and collided with the south China Blocks moved a significant continental mass away (i.e., north) from Pangea's center of mass. This resulted in a pulse of higher surface gravity causing the Triassic-Jurassic extinctions. Terrestrial and marine extinctions were synchronous with the rising levels of gravitation. The resulting regressive/transgressive sea-level couplet caused an extinction of marine lifeforms, during the transgressive phase, that depended on vertical mobility in the water column.

17. Also, as a result of the Cimmerian Plate moving north of the paleoequator, the mantle plume that would eventually manifest itself as the CAMP flood basalt eruptions was initiated by the reversal of the direction of the Earth's inner core movement.

18. The CAMP eruptions, caused by the movement of the core elements toward their geocentric position and the Cimmerian Plate movement initiated the breakup of the Pangean super-continent.

19. The End-Devonian extinctions were caused by the movement of the Gondwana super-continent, first away from the southern pole (lowering surface gravity) and then reversing direction and moving over the southern pole later in the Devonian (increasing surface gravity).

FINAL THOUGHTS

If a dynamic model of the Earth were constructed, displaying the movement of the continental tectonic plates along with the corresponding movement of the Earth's inner/outer cores, it would demonstrate the following:

- During the **end-Ordovician**, the southern super-continent of Gondwana moved over the southern pole, resulting in higher surface gravity on all continental land masses and therefore, extinction.
- During the **late-Devonian**, Gondwana moved away (i.e., north) from the southern pole resulting in lower surface gravity and therefore, extinction primarily of benthic marine life. It then reversed direction and moved south over the southern pole again resulting in increased surface gravity and therefore, extinction. This is why the late-Devonian extinction took place during a much greater time period than the other major mass extinctions.
- The **late-Permian** (i.e., P-T) was a period in which Pangea was at or near final consolidation. It was also moving north toward an equatorial position (increasing surface "g") and the Cimmerian Plate was rotating toward the equator. All three of these actions were increasing surface gravity. The resultant drop in sea levels, along with warm temperatures, caused the

disassociation of methane from the submarine methyl hydrates; the killing mechanism, along with increasing surface gravitation, for the P-T extinctions.

- The **end-Triassic** (i.e., T-J) period was the time during which the breakup of Pangea began. The author of the GTME believes that the pendulum-like movement of the Cimmerian Plate, as it swung north of the paleoequator, was the initiator of the breakup of Pangea. Pangea stopped moving north and started moving south reversing the Earth's inner iron core movement to an Earth-centric direction initiating the CAMP volcanism. The reversal of the inner/outer core direction would also have produced a pulse of increasing surface gravity responsible for the Triassic-Jurassic extinctions. It also caused a regressive/transgressive couplet with its associated transgressive marine extinction.

- The **end-Cretaceous** (i.e., K-T) period was the time when the continental plates, having split apart, began their primarily east-west separation. This would be the time of greatest continuous acceleration of surface gravity. The extinctions of this period have been studied extensively. The timing of these extinctions is in agreement with the GTME:

First, the largest terrestrial animals (well before K-T).

Second, large terrestrial animals and large, fast-moving marine reptiles that were viviparous (i.e., gave live birth).

Third, marine reptiles and shelled marine animals that relied on vertical movement in the water column but were not able to evolve a structure compatible with the effects of increased gravitation.

Fourth, microorganisms with calcareous shells that floated near the surface of the sea and whose buoyancy was compromised by the increased gravitation.

Today, at this very moment, the continents are still moving. They will rendezvous and coalesce again millions of years from now. Surface gravity will become lower than what it is today and some lifeforms will probably become gigantic again. Inevitably, surface gravity will subsequently increase, during which time all existing lifeforms will pass through the "gravity filter" and another extinction will be added to the Big Five family of mass extinctions. Will there be any humans around at that time to record those events? That's a question I don't have an answer for.

The following part of this book contains the additional information of the second edition. As noted in the preface, Part-I through Part-V contain the original edition.

As the GTME evolved, there have been some changes made to the original theory; they will become apparent in the chapters that follow. However, the basic theory has not changed.

166

PART VI: REENFORCEMENT

CHAPTER 17: THE PANGEA A VS. PANGEA B CONTROVERSY SUPPORTS THE GRAVITY THEORY OF MASS EXTINCTION

When Alfred Wegener, in his 1915 book *The Origin of Continents and Oceans*, hypothesized that the continents of the Earth had coalesced to form a super-continent known today as Pangea, he did so without the assistance of the science of paleomagnetism. It wasn't until the 1950s that this branch of science gained traction due to advances in mapping magnetic anomalies on the ocean floor. With the aid of advanced magnetometers the ocean floor was studied and it was found that on either side of the mid-Atlantic Ocean ridge, which extends from high northern latitudes to high southern latitudes, there was a symmetrical magnetic striping of the ocean floor on either side of the ridge. Each stripe signified the direction of Earth's magnetic field at the time that part of the ocean floor was formed and was proof that there was a conveyer belt-type recycling of the ocean floor.

The discovery of the spreading ocean floor was a turning point regarding the prior belief that the continents were immovable, providing overwhelming support for Wegener's ideas. Wegener died prematurely in 1930 while exploring the Greenland ice cap, long before paleomagnetic science confirmed his hypothesis. His Continental Drift Theory, known today simply as Plate Tectonics, was based totally on circumstantial evidence supporting the existence of the Pangean super-continent:

1. The congruence of the Atlantic coastlines of today's continents.
2. The continuity of geologic features that became apparent when the continents were consolidated in a particular way. In the north, mountain ranges such as the Caledonian could be traced from western Europe into Newfoundland and Nova Scotia. In the south, the South African Cape Fold Belt appear to be a continuation of Argentina's Buenos Aires Mountain Province.
3. The similarity of ancient flora and fauna in regions located currently on widely separated continents which would have been

in close proximity if the continents were fitted together. Examples include the fossil remains of tropical plant Glossopteris found only in southern continents and the small Permian reptile Mesosaurus, which was known to inhabit only Brazil and South Africa.

Later confirmation of Wegener's hypothesis of Pangea's structure, roughly as shown in Fig. 17.1, gradually attained acceptance by most of the scientific community. As paleomagnetists accumulated extensive samples of the Earth's crust from continental and oceanic drilled cores, something unexpected and troublesome was discovered, which will be described next.

PANGEA DURING PERMIAN

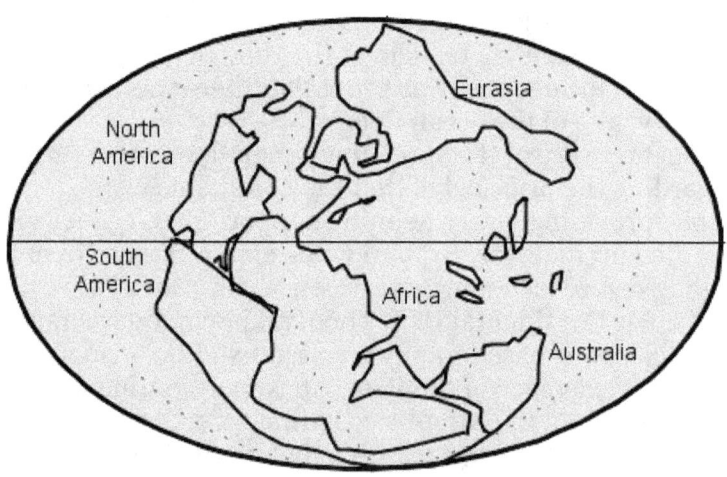

Figure 17.1   Pangea A model based on Wegener

THE GONDWANA/LAURASIA OVERLAP

Paleomagnetists have been guided by the belief that the magnetic field of the Earth is similar to that of a bar magnet lying along the Earth's rotational axis and that even though there can be minor variations in the alignment of the two, over hundreds or thousands of years the magnetic and spin axis coincide when time-averaged over millions of years. In other words, the Earth's magnetic field should always be a viewed as a **Geocentric Axial Dipole (GAD)**. The north pole would be at one end of the central axis, the south pole at the opposite end and a symmetrical magnetic field surrounds the axis as shown in Fig. 17.2. With the GAD model it can be seen that the magnetic inclination angle **I** (the angle between the tangent or horizontal line at any point on the Earth and the direction of the magnetic field) could be determined if the latitude were known. Mathematically, the relationship is:

$$\tan I = 2 \tan L \quad \text{(Where } L \text{ is the latitude)}$$

Using this relationship, paleomagnetists could examine the magnetic inclination of successive layers of crust in a specific location and map the latitudinal (i.e., north/south) location of that part of a landmass going as far back as hundreds of millions of years. This is exactly what paleomagnetists did using a large number of samples and sensitive magnetometer equipment in order to obtain an accurate history of the movement of continental tectonic plates. When they did this, their results indicated that if they went back prior to about 200 mya, their data showed that the southwestern region of Laurasia overlapped the northwestern region of Gondwana. Obviously this could not have happened. The dilemma that was uncovered had to be resolved. The following drawing shows the Geocentric Axial Dipole (GAD) magnetic field of the Earth.

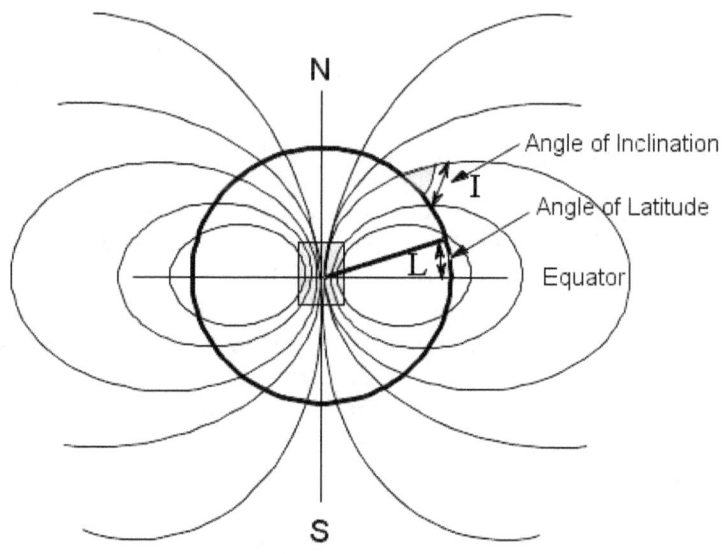

Earth's Dipole

Figure 17.2

The paleomagnetists were using the Wegener model of Pangea (see Fig. 17.3) currently referred to as the **Pangea A** configuration. They were perplexed by the apparent overlapping misfit of the southern and northern continents derived from the paleomagnetic data. Their data from the Jurassic Period (of about 200 my) to the present only roughly agreed with the Pangea A model but paleomagnetic data for older periods indicated their was a Laurasia/Gondwana overlap. In order to resolve this problem they devised a pre-Jurassic model of Pangea which is known as the **Pangea B** configuration and is based on paleomagnetic data. It is shown in Fig. 17.3.

The **Pangea B** model requires that Gondwana (primarily Africa and South America) be shifted east by at least 2000 kilometers (some paleomagnetists claim as much as 3500 kilometers).

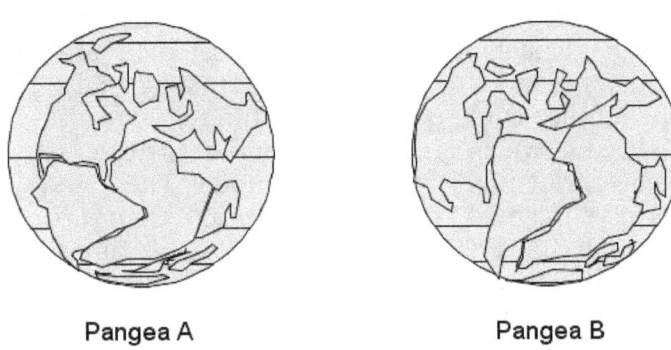

Pangea A   Pangea B

Figure 17.3

Based on the **Pangea B** hypothesis, during an interval of about 20my between the Permian and Jurassic periods, the Gondwana continent moved west by the same 2000 (or greater) km to form the **Pangea A** configuration. This movement is technically known as a dextral megashear between the northern and southern continents and would have left observable indications of this movement. No expected geological remnants of the megashear have been found.

Not only does there appear to be a discrepancy between the Laurasian and Gondwana latitudes based on paleomagnetic data but also within the continental landmass of Laurasia itself there is a discrepancy between its northern and southern regions, an additional problem which has not been solved. To complicate things even further, the predicted magnetic inclinations of Central Asia (e.g., northern Tibet) were found to be much lower than at the corresponding latitudinal European locations.

171

THE PURPORTED NON-DIPOLE MAGNETIC FIELDS

The paleomagnetists who supported the **Pangea A** model now had to explain the apparent conflict between the anomalous paleomagnetic inclination data and their model which was supported by geological, paleoclimate and fossil data rather than accept the existence of **Pangea B**. They did that by analyzing what would happen if they could combine two different types of magnetic fields, known as **quadrapole** and **octupole,** with the standard Geocentric Axial Dipole (GAD) model. Their view was that if the Earth's magnetic field was not a 100% Geocentric Axial Dipole field at all times, then there must have been times when one or both of the above non-dipole fields contributed to the overall magnetic field. Most of their analysis supported an octupole component.

One study, *Evidence for late Paleozoic and Mesozoic non-dipole fields provide an explanation for the Pangea reconstruction problems* by Rob Van der Voo and Trond H. Torsvik, was based on research of two periods, the Late Carboniferous (c. 300 mya) and the Late Permian-Early Triassic (c. 250 mya). They reconstructed the Earth's continents in three ways for both of the two previously named periods:

1. Using the GAD model (i.e., a 100% Geocentric axial dipole magnetic field).
2. Using a 90% GAD and 10% octupole field.
3. Using an 80% GAD and 20% octupole field.

One of the striking conclusions of their study is that if a sufficient octupole component of the Earth's magnetic field were present, the **Pangea A** model can be justified for pre-200myr old periods without the need for the alternate Pangea B:

Late Carboniferous (ca. 300 Mya)
Using the above three reconstruction options, both the second and

third would produce a **Pangea A** super-continent. In other words, either a 10% or 20% octupole component would suffice.

Late Permian-Early Triassic (ca. 250 Mya)
Using the above three reconstruction options, only the third would produce a **Pangea A** super-continent. In other words, a 20% octupole would be required.

The question that arises and doesn't seem to be adequately answered is:
What would cause an octupole component to arise and why would its strength vary for different long-lasting time periods? No acceptable answer has been proffered.

DOES THE GRAVITY THEORY OF MASS EXTINCTION (GTME) SOLVE THE PANGEA A/PANGEA B CONTROVERSY?

The GTME, as explained in most of this book, posits changing surface gravitation based on the movement of the **Earth's core elements** (i.e., inner core, outer core and densest part of lower mantle) from their current geocentric location. The magnitude of this movement is governed by the relative location and consolidation of the continental land masses. As Pangea formed, the core elements moved away from the Earth's center, reaching their zenith of displacement when Pangea was nearly fully consolidated. This would have been during the late Carboniferous (~320 Myr). Therefore, the Earth's magnetic field would have not been a GAD field for most of the distant past, based on the GTME. As shown in Fig. 17.4, the Earth's magnetic field from the Carboniferous through succeeding periods up to the recent geological past experienced a changing **non-geocentric** axial dipolar field except at 65 Myr and 250 Myr. Although obviously not geocentric, the field may have still have been a 100% axial dipole field, as illustrated.

Reenforcement

From Fig. 17.4 it can be seen that Pangea was consolidating during the late Carboniferous period (~320Myr) with the bulk of its mass in the southern hemisphere. This **major southern asymmetry**, relative to the equator, would have caused the Earth's core elements to be displaced far from their current geocentric position based on the Law of Conservation of Angular Momentum explained in *Chapter 18*. This would cause analysis of paleomagnetic material from that era to be problematic because a GAD assumption was used.

At the end of the Carboniferous, Pangea began to move north at a rapid rate (see Fig. 20.1) and eventually (about 250myr) was positioned so that its center of mass was on the equator. Therefore, based on the GTME, the Earth's core elements had returned to their current geocentric position. Paleomagnetic data from ~250 myr should indicate a GAD magnetic field, and they do:

The following quote is from scientific paper entitled *Early Permian Pangea 'B' to Late Permian 'A'*:

"The transformation from Pangea 'B' to Pangea 'A' took place during the Permian because Late Permian paleomagnetic data allow a Pangea 'A' configuration."

Although I disagree with the cited study's support for the Pangea 'B' model, the authors support what the GTME concludes, which is that the movement of the core elements from far off-center back to their geocentric position simulates the apparent (but false) non-100% dipolar to the 100% dipolar transition.

Therefore, the paleomagnetists' perceived reconstruction problems of Pangea during the Carboniferous through Jurassic Periods (based on the GAD model) disappear when viewed from the GTME's non-geocentric dipole model.

174

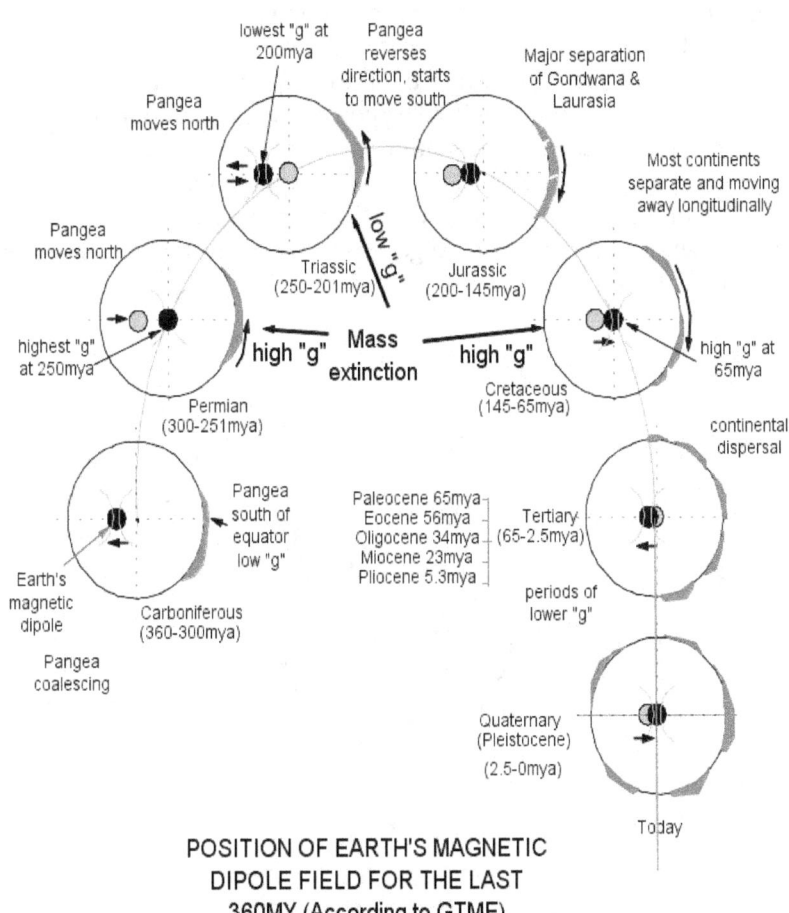

POSITION OF EARTH'S MAGNETIC
DIPOLE FIELD FOR THE LAST
360MY (According to GTME)

Figure 17.4

175

Reenforcement

Figure 17.5 illustrates the displaced axial magnetic dipole that is based on the GTME. As shown, the angle of inclination (**I**) at any latitude (**L**) on Pangea will vary based upon the degree of displacement (i.e., offset) of the core elements of the Earth. As the core elements (and dipole) move further away from the Earth's geocenter the inclination angle gets smaller for a specific latitude of Pangea. Paleomagnetic recordings of the Earth's inclination presents a problem to paleomagnetists only because they have assumed that the core elements of the Earth have always been fixed geocentrically and, it follows, that the magnetic field must also be geocentric. The dashed circle represents the current Earth.

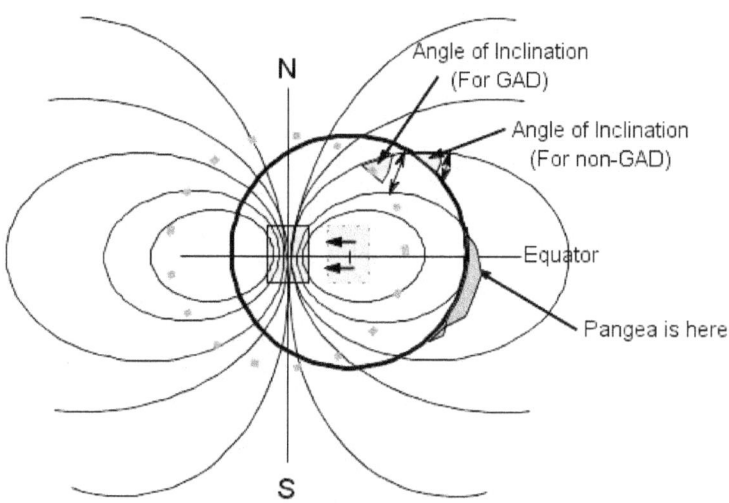

Non-GAD Earth dipole (Based on GTME)
For any latitude, the Inclination Angle would be
lower when Core(s) shifted due to non-GAD.

GAD is Geocentric Axial Dipole

Figure 17.5

Another reference that supports the GTME in relation to the **Pangea A** vs. **Pangea B** controversy comes from *Permanent aspects of the earth's non-dipole magnetic field over upper Tertiary times* by R.L. Wilson. In his publication he suggests that the model that best fits the Tertiary time-averaged geomagnetic field was that of an **"offset" dipole** displaced some 191 $^+_-$ 38 kilometers northward along the axis of rotation. This configuration would simulate the displaced axial dipole hypothesized by the GTME where the core elements are displaced away from Pangea (see Fig. 18.2). Of course, the GTME's "offset" dipole is offset perpendicular to the Earth's rotational axis.

Let's take a closer look at the problem of the apparent overlap of Gondwana and Laurasia based on paleomagnetic analysis. In Fig. 17.5, the dashed circle represents the current Earth with its geocentric axial dipole magnetic field and the solid line circle represents the Earth with the displaced core elements. It can be seen that any location that was not directly on the equator when its paleomagnetic imprint was created will present a problem when the Earth's core elements return to their geocentric position. Its **angle of inclination will be smaller** when the core elements are offset. Assuming it remained at the same latitude when the core elements returned to their geocentric position, paleomagnetists assuming the GAD model would conclude that the location was closer to the equator when its paleomagnetic imprint was formed. This explains the Laurasia/Gondwana overlap problem.

The paleomagnetic discrepancy problem within Laurasia itself, alluded to earlier, is also explained by the above analysis because the flattening of inclination angle is more severe the further the location was from the equator, i.e., the inclination angle stays the same at the equator even when the core elements are displaced. This is apparent from Fig. 17.5.

CHAPTER 18: PHYSICS OF THE GTME

The concept of the movement of the Earth's core elements away from their geocentric position is alien to almost everyone, including those who are knowledgeable when it comes to science. This was proven in the prior chapter where paleomagnetists, trying to reconcile the Pangea A and Pangea B problem, did not deviate from their basic belief that the earth's magnetic field could not be anything but a geocentric one, albeit one with contributions from non-dipole magnetic fields.

In order for the Earth's core elements to shift away from the Earth's geocentric position, there have to be forces acting on those elements to cause that movement. We know, without a shadow of doubt, that the Earth's continents have "drifted" across the globe in many difference configurations. We also know, without a shadow of doubt, that these continents coalesced (about 300 myr to 250 myr) to form a super-continent known as Pangea, or Pangaea inthe southern hemisphere. This super-continent was surrounded by a vast body of water known as the Panthalassa Ocean. The primary question that has to be answered is:

DID THE MOVEMENT OF CONTINENTS DURING THE FORMATION OF PANGEA PRODUCE FORCES WHICH MOVED THE EARTH'S CORE ELEMENTS IN A DIRECTION OPPOSITE TO THAT OF PANGEA?

If the GTME is valid, then the answer to the above question must be "Yes." However, it must be pointed out that since the Earth's continents have been in various configurations throughout the past, there may have been a prior time when the Earth's core elements have been geocentric, as they are today due to a relatively symmetrical distribution of land masses around the globe. During the consolidation of Pangea, there was no recent "starting point" where the Earth's core elements were geocentric. Therefore, it makes sense to analyze what would happen if we start with today's Earth configuration and proceed in reverse toward that of Pangea.

In order to confirm the GTME, the basis of which is the movement of the core elements away from their current geocentric location, no laws of physics can be broken. There are two primary laws that must be addressed: The law of Conservation of Rotational Kinetic Energy and The law of Conservation of Angular Momentum .

CONSERVATION OF ROTATIONAL KINETIC ENERGY

The kinetic energy of a particle moving at a constant velocity is

$$K = 1/2(mv^2)$$ where m is the particle mass
v is the particle velocity

The above equation applies to what is called translational kinetic energy (i.e., the kinetic energy of a particle moving in a straight line). Since we are concerned with the Earth, all of the particles of which are moving in a circular motion, the above equation has to be revised to include rotational parameters. The magnitude of the velocity of a particle in a rotating body is:

$$v = rw$$ where **r** is the distance from the
particle to the rotation axis
w is the angular velocity of the
rotating body

The rotational kinetic energy of a single particle in a rotating body is then, based on the translational kinetic energy:

$$K_r = 1/2(mr^2w^2)$$

Because each particle in a rotating body has a radius (i.e., distance from the axis of rotation) different from most other particles but the same angular velocity, the total rotational kinetic energy for the rotating body is:

$$K_{rtot} = 1/2(w^2)[\text{Sum of } mr^2 \text{ for all particles}]$$

179

Reenforcement

The sum of **mr²** for all particles in a rotating body is defined as the moment of inertia (**I**) of the body. It is a scalar quantity (i.e., it has magnitude but no direction) and is the analogue of mass in translational motion. The total rotational kinetic energy is:

$$K_{rtot} = 1/2 I(w^2)$$

If we consider the Earth, a rotating body, its rotational kinetic energy should remain unchanged unless some external force acts on it, such as the impact of a large meteorite. If the Earth's angular velocity (**w**) is constant (i.e., the length of a day doesn't vary), then the only thing that can significantly change the Earth's rotational kinetic energy (**K**$_{rtot}$) is a change in the moment of inertia (**I**). Since **I** is the sum of **mr²** for all particles in the body and the total mass of all the particles can't change then the only way the Earth's rotational kinetic energy can change significantly, without external influence, is by shifting of mass somewhere within the Earth that causes a change in moment of inertia. Fig. 18.1 is a drawing of the Earth showing the movement of a large continent, such as Australia, that moves to a higher southern latitude.

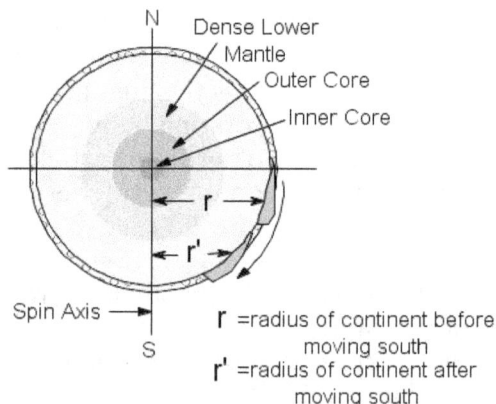

EARTH WITH CONTINENTAL
MOVEMENT

Figure 18.1

The movement of the continent from the initial position with a radius (i.e., distance from spin axis) of **r** to the new position with a smaller radius (**r'**) could not occur without some other offsetting action that would maintain the same rotational kinetic energy of the Earth. The obvious one would be an increase in the rotational velocity (**w**) of the Earth.

If we recreate Pangea from the current Earth configuration we find that the most massive continents initially consolidated themselves south of the equator in a "C" shaped formation. Based on the previous described relationship between the movement of continental mass and conservation of rotational kinetic energy, we would expect that the large decrease in moment of inertia resulting from the continental masses moving south and away from the equator would result in an increase in the Earth's angular velocity. However, to my knowledge, there was no change in the length of a day when Pangea formed and if true, no change in angular velocity. Fig. 18.2 is an illustration of continental position about 300mya.

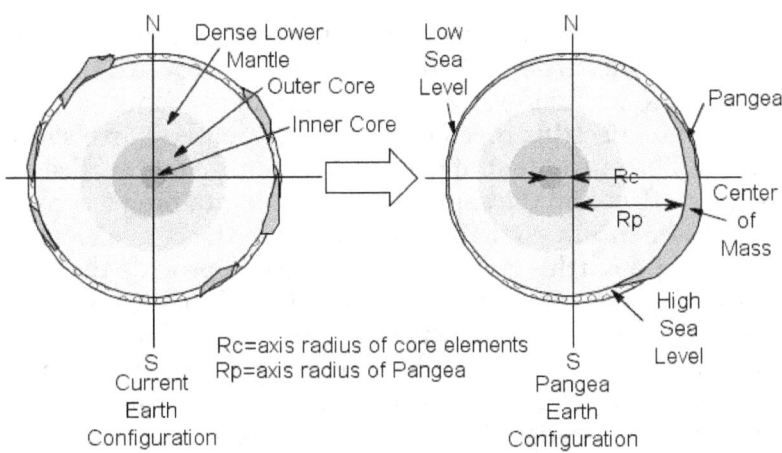

EARTH'S CORE ELEMENTS SHIFT
AS PANGEA IS FORMED
Figure 18.2

Reenforcement

If there were no change in the Earth's angular velocity then there had to be another mechanism to maintain the rotational kinetic energy. The only other one is the movement of some or all of the core elements and their movement would have to be away from their geocentric position, which would increase rotational kinetic energy, because they are more massive, offsetting that lost by the massive continental movement toward the south pole.

CONSERVATION OF ANGULAR MOMENTUM

The Earth's angular momentum, like rotational kinetic energy, must be conserved if no external forces (i.e. torques) act on the Earth. Unlike kinetic energy, angular momentum is not a scalar quantity. It is a vector quantity, meaning that it not only has magnitude but also direction. For a rigid body, angular momentum (**L**) is related to angular velocity and moment of inertia:

$$L = I\omega$$

Since we are assuming that the Earth's rotational velocity did not change when Pangea formed, the only variable that could change the angular momentum is a change to the moment of inertia. As detailed in the prior section, movement of massive continental plates from near the equator south toward the southern pole would lower the Earth's moment of inertia unless there was an offsetting mechanism that increased Earth's moment of inertia. The movement of the core elements could suffice to satisfy the offset and the same is true for maintaining a constant angular momentum. Since the movement of the core elements away from the Earth's geometric center could provide the offsetting angular momentum and rotational kinetic energy, the final problem is to find out why the core elements moved away from Pangea and not towards it since movement away from the geocenter in several directions could offset angular momentum and moment of inertia. The GTME only makes sense if the core elements moved away from Pangea because that would lessen the surface gravity on Pangea.

When an object with mass symmetry about a specific axis is spinning so that the axis of rotation coincides with that axis and there are no external torques acting on it, it is in what is called a stable rotation of minimum energy configuration. If the mass symmetry about the axis is gradually changed by redistributing the mass, the spinning object will start to wobble. It will no longer be in a minimum energy configuration. If the spinning object is constructed of deformable material, the wobble will be dampened as energy is transferred to the deformable material, possibly raising the temperature of the material. This is one mechanism that is used in space satellites to eliminate wobbling and control the attitude (i.e., direction) that the spacecraft is oriented toward, something that is important when cameras or radio transmitters are present. This is known as nutation damping. The same process applies to the Earth. If a mass redistribution causes a wobble, the Earth's deformable interior will dampen the wobble either by raising the internal temperature or doing work to redistribute mass within the Earth, or both. The following quote is from the publication *Geometrical Approach to Torque Free Motion of a Rigid Body Having Internal Energy Dissipation* by Philippe L. Lamy and Joseph A. Burns:

"Thus spinning bodies with any internal damping mechanism, e.g., any elastic imperfections, will always be moving towards aligning themselves in their minimum energy configuration, i.e., in pure rotation about their principal axis of maximum inertia."

Based on the conservation of kinetic energy and angular momentum, it is clear that the movement of the massive continental plates to higher southern latitudes required a corresponding movement of the core elements, in whole or in part, away from their geocentric position. And, based on the minimum energy configuration principle and maximum inertia principal axis cited above, the core elements would have to have moved radially away from the center of mass of Pangea as shown in Fig. 18.2 to dampen the expected wobble.

Reenforcement

SURFACE GRAVITY CHANGE ON PANGEA

The Gravity Theory of Mass Extinction (**GTME**) is based on the premise that the movement of the Earth's core elements throughout the past have altered the surface gravity on land masses resulting in extinction episodes. Is it possible to know the magnitude of the changes in surface gravity with any degree of accuracy? Probably not. In the case of Pangea, one might be able to infer a range of values for surface gravity based on the magnitude of paleomagnetic discrepancies described earlier.

      Using the basic Newtonian gravity laws, we can estimate how much the core elements had to shift from their geocentric position to produce a specific change in surface gravity (i.e., "g"). The following assumptions will be used to do that:

1. The Earth's inner iron core represents 2% of the Earth's mass.
2. The Earth's outer liquid core represents 31% of the Earth's mass.
3. The Earth's dense lower mantle represents 52% of the Earth's mass.
4. All of the above three core elements moved symmetrically away from center. Elements 2 & 3 above most likely deformed from their spherical shape when they came under non-symmetrical centrifugal forces but for simplicity, let's assume they didn't.

A very rough estimate of the displacement of the core elements can be made. Referring to Fig. 18.3, today's weight of an object, let's say at the equator, can be compared to when Pangea existed. Today's weight is $\mathbf{W_T}$ and Pangea's weight would be $\mathbf{W_P}$. The weight at the surface of the Earth today for an object of mass **m** would be:

$$\mathbf{W_T} = GmM_E/R_E{}^2$$

    Where G is the Gravitational Constant
                 M_E is the mass of the entire Earth
                 R_E is the distance from the object to
                     the center of mass of the Earth.

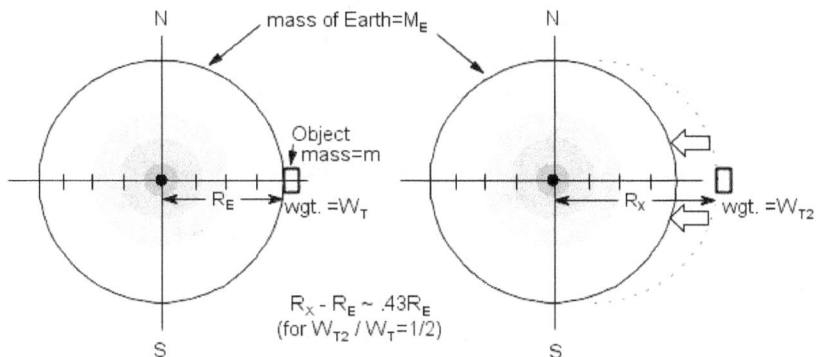

CHANGE IN WEIGHT OF OBJECT BASED ON
DISTANCE FROM EARTH'S CENTER OF MASS

Figure 18.3

If the object could remain stationary and the Earth moved away from it, its weight would decrease. If we want to know at what distance the Earth would have to move away in order for the object to weigh one half of what it did at the surface, calling the new weight W_{T2}), then:

$W_{T2}/W_T = \frac{1}{2} = (GmM_E/R_X^2)/(GmM_E/R_E^2)$

Where R_X = the distance from the object to the center of the mass of the Earth when its weight is halved. Cancelling out the constant terms results in:

$W_{T2}/W_T = \frac{1}{2} = R_E^2/R_X^2$ or

R_X = square root of 2 R_E^2

In Fig. 18.3, the Earth's diameter has been divided into 8 equal units. To get an idea of the relative amount of movement of the Earth to halve the object's weight in the above example:

R_X = square root of 2 x 16 ~ 5.7 units

Reenforcement

The Earth would have to move away from the object by $R_X - R_E$ or 5.7 - 4.0 = 1.7 units. Since the radius of the Earth is 4 units, movement of the Earth by less than half (about 43%) of its radius would be needed.

Instead of having the object stationary and the Earth moving away, the same calculation can be done using both a stationary object and Earth with the Earth's core elements moving away. With the core elements having a mass approximately 85% of the Earth, the distance the core elements would need to move (R_X), would have to be somewhat longer. Instead of 43% of the Earth's radius, estimating 50% of the radius would not be unreasonable. Therefore, for $W_P = \frac{1}{2}W_T$, the center of mass of the Earth's core elements would have to be displaced about 1/4 of the diameter of the Earth.

Some of the assumptions that the calculation was based on were listed above. Also, the spherical symmetry of the core elements might start to lessen as they moved away from their Earth-centric position. Centrifugal forces might have flattened, or widened, the outer core and dense lower mantle elements as they moved off-center and the inner core may well have had its angular velocity altered.

In conclusion, if an object at Pangea's center of mass (i.e., place of lowest "g") weighed one-half of what it would on today's Earth, the center of mass of the core elements would have to have shifted away from the geocentric center of the Earth by about half the Earth's radius (i.e., one quarter of its diameter).

.When Pangea moved north during the Permian Period, its center of mass moved across the equator, increasing its surface gravity to very high values and causing one of the greatest mass extinctions of all time.

186

CHAPTER 19: LOOSE ENDS

There are a few more bits of circumstantial evidence that support the GTME although I believe the Pangea A vs. Pangea B controversy explained in *Chapter 17* is the most powerful. The following topics describe some of this support:

THE CRETACEOUS AND KIAMAN SUPERCHRONS

The reversal of the Earth's magnetic north and south poles occurs approximately every 100,000 to 1,000,000 years. Yet, there were two instances when the reversal didn't happen for many millions of years; these periods are known as **superchrons**. The **Cretaceous Normal Superchron** (120-83 myr) lasted 37 my and the **Kiaman Reverse Superchron** (312-262myr) lasted 50 my. Earth scientists offer no convincing explanation for this anomaly. Since the Earth's magnetic field is generated by the convective movement of the outer core's molten iron around the solid inner core, something unusual must have been happening relative to these two core elements during the time the superchrons existed.

What is coincidental about the two superchrons is that they occurred during the two significant periods of the history of Pangea. The Kiaman Reverse Superchron occurred as Pangea was slowly consolidating to its final state. The Cretaceous Normal Superchron occurred when Pangea's continents had broken apart and started dispersing. The duration of both superchrons parallels the slow formation/more rapid dispersal movement of Pangea.

The GTME posits that the Earth's core elements were moving away from the geocenter during the time of the Kiaman Reverse Superchron and moving toward the geocenter during the time of the Cretaceous Normal Superchron. Another possibility is that the core elements were static during these two periods. Therefore several questions have to be raised:

1. Is it a coincidence the Kiaman Reverse Superchron occurred during the period when the GTME posits that the Earth's core elements were moving away from the geocenter nearing the zenith of their displacement?

Reenforcement

2. Is it a coincidence that the Cretaceous Normal Superchron occurred during the period when the GTME posits that the Earth's core elements were rapidly moving toward their current geocentric position?
3. Is it a coincidence that the two superchrons have reversed polarities when the GTME posits that the Earth's core elements were moving in opposite directions during the two superchrons?

FLOOD BASALT VOLCANISM

As noted in an earlier chapter, there have been episodes of eruptions of flood basalt volcanism (sometimes called hot-spot volcanos) and the most massive of these occurred between about 260 to 65 myr. They have continued since that period but their intensity or outpouring volume has gradually diminished over time. This special type of volcanism is believed to originate at the Earth's core/mantle boundary producing plumes that, some say, take perhaps up to a million years to rise to the Earth's surface. The Siberian Traps (ca. 251 myr) and the Deccan Traps (ca. 65 myr) are well known examples.

The GTME attributes this type of volcanism to translational movement of one or more of the core elements and indirectly by extension, based on the theory, from continental plate movement. An example to support this is the massive CAMP (Central Atlantic Magmatic Province) flood basalt eruption of about 200 myr. This is the period when the continental separation that formed the incipient Atlantic Ocean occurred. There is a question of which occurred first, the continental separation or the volcanic eruption; many believe that latter came first. The GTME would sequence the events in the following order: continental separation, which caused core element movement, which initiated the core/mantle plume. The period was one of the 5 largest mass extinctions where both substantial flora and fauna disappeared. No climate change has been discovered for the period nor bolide impact. One factor that paleontologists discovered was that there was a rapid sea level regression immediately followed by transgression. The sea-level drop appears to be the only reasonable cause of the extinctions (according to paleontologists) but the extinction of terrestrial flora and fauna throws a wrench into that assumption.

The end-Triassic extinctions can be explained using the GTME. Summarizing the events in more detail using the theory, the following events would have occurred in the this order:

1. Major continental separation between Gondwana and Laurasia in the region between northwest Africa and northeast USA creates the incipient Atlantic Ocean.

2. The continental separation, moving large masses latitudinally, causes the Earth's core elements to move from their substantially displaced position back toward the geocentric position as detailed in *Chapter 18*. This would also cause a pulse of increasing surface gravitation on Pangea. The gravity gradient on Pangea, as described in an earlier chapter, produced the lowest surface gravity near Pangea's center of mass, which would have been well near the present equator; the region where changes in "g" would be the greatest.

3. The pulse of increasing gravity would have affected both terrestrial flora and fauna as well as marine life. We know there was extinction in all three of those categories.

4. The movement of the core elements would have initiated a plume of molten material at the core/mantle boundary which would have taken at least a million years to reach the surface. In *Mass Extinctions And Their Aftermath* by A. Hallam and P.B. Wignall, the authors, discounting volcanism as the cause of the extinctions, state: "Therefore the eruptions took place several million years after the extinctions, clearly ruling them out as a possible causal factor." This statement supports the GTME assertion that it was a gravitational change, not the delayed volcanism, that caused the extinctions.

5. The pulse of gravitational increase on Pangea would have created a rapid lowering of sea level (i.e., regression), most intense near Pangea's center of mass due to the gravity gradient previously described and can be seen in Fig. 18.2. It would have been followed rapidly (in geologic time scale) by a rebounding transgression. The higher felt- effects of the changing sea levels would be near Pangea's center of mass, as would be expected from the GTME, is supported by Hallam and Wignall in the book cited above by their statement "A glance at Fig. 6.8 shows that all these regions are located centrally to Pangaea, and it may not be coincidental that the regression-transgression couplet is most marked here; it could possibly relate to the major tensional event causing rifting and volcanicity in the southern North Atlantic region......."

The above conditions of the end-Triassic period parallel those at end-Cretaceous, the K-T Extinctions. The major difference is the bolide impact at Chicxulub. Due to the coincidence in time of the elevated extinction rates and the impact, there is widespread belief that the impact was the major cause of the K-T extinctions. The GTME is able to account for the extinctions, the Deccan Traps volcanism and the regression/transgression couplet that are coincident at the boundary. Logically, the question that arises is:

Is it a complete coincidence that the Chicxulub impact happened at the K-T boundary when major extinctions were occurring?

The GTME has an answer for that question. When the continental masses were rapidly separating during the end Cretaceous period, it would be expected that the Earth would develop a wobble that would eventually be dampened by the movement of the core elements as they moved back towards their current geocentric position. This would be a substantial wobble compared to anything observed in modern times. The spinning Earth, with this extreme wobble, would disturb the path of nearby celestial bodies, whether they be planets, asteroids or comets. It would also present a much wider target for comets and asteroids to collide with, and it would not be unreasonable to assume that widest part of the Earth (i.e., nearest the equator) would most likely be the recipient of the impact.

The asteroid or comet that struck the Earth around 65 mya did impact at a near-equator location, the Yucatan Peninsula in Mexico! Is there any proof that the Earth had an unusual wobble at the K-T boundary? Yes, an article published in CNN Tech *Earth-shaking 'wobble' may have killed dinosaurs* by Thom Patterson describes how scientists at the Univ. of California at Los Angeles determined that there was a wobble in Earth's and Mercury's orbit during the Cretaceous. They believe the gravitational effects of the wobble of the two planets may have caused an asteroid to break away from the asteroid belt between Mars and Jupiter which then went on to collide with the Earth. They are not sure if it was a comet or an asteroid. While the wobble discovery indirectly supports the GTME, I believe that the K-T extinctions, like the end-Triassic extinctions, were primarily caused by changes in surface gravity, not by impact nor volcanism, as described throughout this book.

Conclusion

1. The most powerful circumstantial evidence supporting the GTME is the Pangea A vs. Pangea B controversy.

2. The laws of physics are not violated by the GTME. The consolidation of Pangea moved the total continental center of mass to a higher southern latitude (compared to today), which would have lowered the Earth's angular momentum if not offset by either an increase in the Earth's rotational velocity or core element movement. Higher rotational velocity (i.e., a shorter day) has not been detected during the existence of Pangea, leaving the movement of the Earth's core elements as the only other offsetting mechanism.

3. Based on 1 & 2 above, movement of the Earth's core elements which would alter surface gravitation, establishes the GTME as a bona fide theory that accounts for many, if not most of the Earth's mass extinctions.

CHAPTER 20: CONFIRMATION OF THE GTME

Recent Hypothesis Links Plate Tectonics and Geomagnetic Reversals.

A recent study published (on October 11, 2011) in *GEOPHYSICAL RESEARCH LETTERS* hypothesizes that plate tectonics is the driving force that determines the frequency rate of Earth's geomagnetic field reversals. The study (which I will refer to as the Plate/Reversal Hypothesis) is entitled *Plate tectonics may control geomagnetic reversal frequency*; the authors are F. Petrelis, J. Besse and J. Valet.

This study hypothesizes that there is a direct correlation between the frequency of the Earth's geomagnetic field reversals and the positioning of the Earth's continental plates. More specifically, the authors assert that in geological intervals during which there is an asymmetrical distribution of the continents relative to the equator, the intervals that follow are characterized by high reversal frequency.

The GTME, as stated in earlier editions of this book which were published well before the study cited above, suggested the linkage between continental plate movement and geomagnetic reversals with regard to the Kiaman superchron and the Cretaceous superchron. A detailed comparison of the Plate /Reversal Hypothesis and the GTME concerning geomagnetic field reversals will be described in a subsequent section.

The significance of the above study is the disclosure of the detailed graphical data illustrating how the continental plate distribution relative to the equator changed over the last 300 Myr. As the author of the GTME, I was not aware of the shape of the curve revealed in the published study, although I knew that the supercontinent of Pangea did move north and south during the Paleozoic and Mesozoic Eras. The graph of the study has been redrawn below with commentary added. It has provided major confirmatory support for the GTME, **the basis of which links continental plate distribution and core element movement.**

192

A Major Confirmation of the GTME

Chapter 15 of this book explains how eustatic sea levels would be affected by changes in the Earth's surface gravity, according to the GTME. Recall that as the core elements (i.e., inner core, outer core and densest part of lower mantle) moved away from their current geocentric position and away from Pangea, a surface gravitational gradient would have formed; surface gravity lowest near Pangea's central equatorial region and highest antipodally (which would have been within the vast Panthalassa Ocean.

    Chapter 18 explains the linkage between the latitudinal position of the center of mass of Pangea and the displacement of the core elements (which altered surface gravity.) This linkage is based on the Law of Conservation of Angular Momentum. The further the center of mass of Pangea moved from the paleoequator, the lower the surface gravity on Pangea. Eustatic sea levels would have a depth gradient similar to surface gravity but in reverse; highest sea levels where surface gravity was lowest (i.e., near Pangea's center of mass) and lowest sea levels where surface gravity were highest.

    The Plate/Reversal Hypothesis cited in the prior section strongly, but indirectly, supports the above correlation between surface gravity levels and sea levels during the past 320 Myr as posited by the GTME. This will become apparent by viewing Fig. 20.1 which compares the magnitude of sea levels (using data from the Hallam et al. sea level curve) and the continental plate symmetry curve provided by the above cited Plate/Reversal Hypothesis study.

    Note that the continental plate symmetry graph displays two curves. The data used for the solid line curve was supplied by the researchers of the study and the dash line curve is from data supplied by C. Scotese. Christopher Scotese has produced many excellent illustrations of Pangea on his website www.scotese.com.

    The '**D**' parameter in the graph is a quantification of the continental plate area asymmetry relative to the paleoequator.

Reenforcement

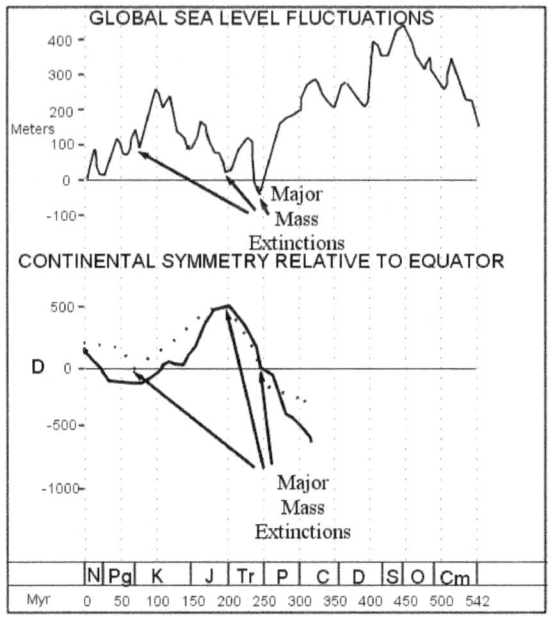

Figure 20.1 see text for source of data

Comparing the sea level curve to the continental plate symmetry curves, it is apparent that there were two periods in the last 320 Myr when there was a rapid and significant drop in sea level, roughly at 65 Myr and 250 Myr. The continental plate symmetry curves indicate these same two periods have maximum symmetry relative to the equator (i.e, the curves cross the zero '**D**' line) although around 65mya the two continental plate symmetry curves differ somewhat. The Scotese curve, in my opinion, supports the GTME more fully at the 65 Myr reference point.

According to the GTME, as was described in an earlier paragraph, these two periods would be periods of relative high surface gravity on Pangea because the core elements would have returned to near their current geocentric position. The GTME would interpret the continental symmetry graph as follows:

194

When the continental plate symmetry curve is near the zero 'D' line (indicating high symmetry) surface gravity was near its highest level on Pangea.

The drop in sea levels at 65Myr and 250 Myr shown on the sea level curve coincides with the two periods during which the GTME would expect surface gravity to be high based on the high symmetry indicated in the continental plate symmetry curves, and this is what we observe in Fig. 20.1. This is major supporting evidence for the confirmation of the GTME. And, even more important are the well known mass extinctions during these two periods. Therefore, there would have been a coincidence of high surface gravity, extinction and low sea levels during these intervals, which is precisely what the GTME posits. The Plate/Reversal Hypothesis has played an important role in validating the GTME.

If the relationship between sea level change, extinction and continental plate symmetry (and by extension surface gravity per the GTME) is true in general, there should be other periods when the three conditions described in the previous paragraph coincided. The Triassic/Jurassic extinction occurred about 200 Myr. The sea level graph indicates a rapid drop in sea level at that time. The continental plate symmetry graph indicates that surface gravity was being lowered to near its lowest value during the last 320 myr but abruptly (geologically speaking) reversed and surface gravity started to increase. This would have caused a pulse of increasing surface gravity and a corresponding extinction event. The GTME posits that the splayed- leg crurotarsans went extinct at this time because their leg structure would not support their great weight with the rapidly increasing surface gravity, yet dinosaurs with vertical leg structure would be less affected. The crurotarsans were the primary competitors of the dinosaurs at that time and their extinction opened the door for the reign of the dinosaurs.

Reenforcement

Mass extinction at 200 Myr has been well documented.

The continental plate symmetry graph is not sufficiently detailed to identify all extinction periods during the last 320 Myr. However, the sea level graph does display many drops in sea level, both large and small, and the GTME will make the following assertion:

The rapid drops in sea level as indicated by the sea level curve (from Hallam et al.) coincide with mass extinction events caused by pulses of increasing surface gravity possibly with the exception of any periods of large glacial or polar ice formation.

It has to be noted that the '**D**' parameter in the continental symmetry graph is a measure of symmetry of the total surface area of the continents relative to the paleoequator. The GTME would use a different definition of the '**D**' parameter as described in the following paragraphs. The different '**D**' parameter would take into consideration the moment of inertia of each continent rather than just surface area. The moment of inertia of a continent is determined by its latitude and by the mass of the continent and how it is distributed vertically, i.e., the distance of the mass from the Earth's rotational axis. For the purpose of illustration, assume that a continent is circular and has a large, tall mountain range localized on its periphery. If that continent rotated but remained at the same latitude, then the '**D**' parameter used by the Plate/Reversal Hypothesis would not change yet the different '**D**' parameter of the GTME would change as a result of the continents change in moment of inertia, i.e., the mountain range would have moved to a different latitude.

Based on the above, the continental plate symmetry curve of the Plate/Reversal Hypothesis would be similar to, but somewhat different than the curve that would more accurately describe changes in surface gravity and geomagnetic field

196

reversals as well. However, as published, the continental plate graph is good news for the GTME; it provides major support for the gravity theory.

One can find many graphs on the internet that display the coincidence of extinction periods and episodes of flood basalt volcanism; the Deccan Traps and Siberian Traps are the major examples found. Many have assumed that the volcanism caused the extinctions because of the timing coincidence. As explained in prior editions of this book, the GTME believes the flood basalt volcanic eruptions were caused by the higher than normal fluid pressure within the outer core as the core elements were moving back toward their current geocentric position. I call this the "plunger effect." If the GTME is correct, the drops in sea level on the Hallam et al. curve should coincide with major flood basalt eruptions. I haven't checked all of the matching sea level dips and flood basalt eruptions but the pattern seems to conform to the one expected. Therefore, GTME makes the following assertion, which was made early in the evolution of this theory:

> **Massive flood basalt volcanism is the result of core element(s) movement toward the Earth's geocentric position and therefore follow pulses of increasing surface gravity, which caused the preceding extinction.**

Note that the flood basalt eruptions may take place over millions of years. Therefore the corresponding extinction pulses would precede and might recur during the eruptions. The magnitude and timing of the successive flows of lava at the surface might reflect the intensity of the pulses of increasing surface gravity, and therefore, the intensity of the extinction pulses.

Reenforcement

Geomagnetic Reversals

The primary conclusion of the Plate/Reversal Hypothesis is that plate tectonics is directly linked to the frequency of the Earth's geomagnetic field reversals. The authors of the study do not give an indisputable cause-and-effect mechanism to account for this linkage. One possibility, they offer, is that slabs of oceanic crust are subducted and eventually descend to the core/mantle boundary (CMB) where they cause "thermal heterogeneities in the lower mantle." This process would, according to the authors, "imply an offset of 40-50 Myr between changes at the CMB and plate motions at the surface."

The GTME would explain the plate movement/ geomagnetic reversals in a different way: Continental plate movement is directly linked to the movement of one or more of the core elements; they move unison. The plate positions are linked to the core elements based on the Conservation of Angular Momentum Law. However, the geomagnetic reversals might lag the plate movement but it would by time intervals much shorter than that proposed by the Plate/Reversal Hypothesis and the lag would depend on the mass, velocity and direction of the plate movement. Therefore the GTME asserts:

The Earth's geomagnetic field reversal rate depends on the spherical symmetry of the inner core within the outer core, and, that symmetry is governed by continental plate position. The greater the asymmetry, the more likely a reversal will occur.

Once the core elements move away from the Earth's geocenter, the inner core/outer core symmetry can vary significantly. When Pangea consolidated into a supercontinent the potential changes to the angular momentum of the Earth would have been very large, this is why all three of the core elements (inner and outer core and densest part of lower mantle) had to move off center to compensate primarily for the latitudinal movement of the supercontinent. Today, with the continents dispersed uniformly around the globe, only the inner core needs to move within the outer core to compensate for continental movement. And, some continental movement could be cancelled by other continental movement. This would happen for example if one continent moves south toward the equator and a similar sized continent in the same hemisphere moved north away from the equator. Also, today the inner core can move away from spherical symmetry within the outer core in the full 360 degrees of its equatorial plane, whereas during Pangea existence it had to move away from Pangea. This is illustrated in Fig. 20.2.

MOVEMENT OF INNER CORE WITHIN OUTER CORE

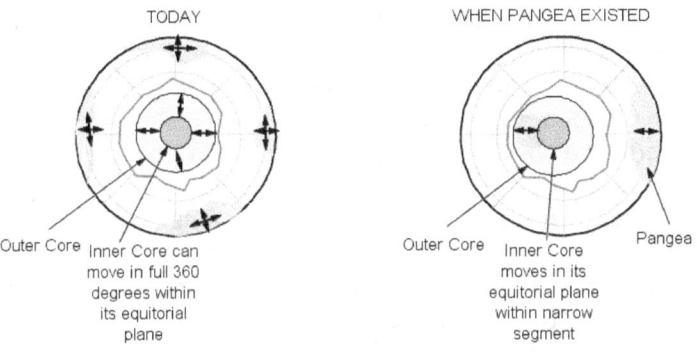

Figure 20.2

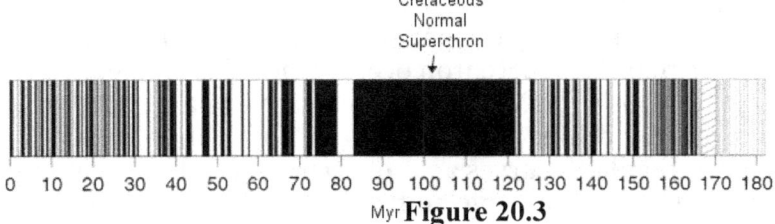

Myr **Figure 20.3**

199

Reenforcement

Figure 20.3 is a chart of the Earth's geomagnetic reversals for the past approximately 180 Myr. The Cretaceous Normal superchron is clearly identifiable near the center of the chart. One characteristic of the reversal pattern is the duration of each period, or chron. During the past 30 Myr the duration of each period is rather brief, probably much less than one million years while the duration is much longer near the superchron. Viewing Figures 20.2 and 20.3 together, the GTME might explain this variation in chron duration as follows:

As explained in a prior paragraph, the continents today are uniformly distributed around the globe. This means that each continental movement may cause a slight displacement of the inner core in any of 360 degree possible ways along the equatorial plane of the inner core. If, for example, a single large continent moves latitudinally, it may displace the inner core sufficiently to disrupt the symmetric flow of the outer core's molten iron and initiated a magnetic reversal. With the continents dispersed the way they are today, the movement of single continents can result in a higher frequency of reversals.

When the continents were amassed on one side of the globe and dispersing, primarily longitudinally, away from Pangea, the inner core's movement would be limited to a narrow pie-shaped segment of its equatorial plane, unlike today. This would lengthen the frequency of reversal. Most likely the reversals during this period were initiated by latitudinal movement or rotation of continents as they were dispersing from the consolidated Pangea.

In the case of superchrons, the inner core is able to maintain near 100% symmetry within the outer core for very long periods of time. Based on the continental plate/reversal graph it appears that this can only happen when a supercontinent exists. It also appears that the superchron occurs when the supercontinent is moving towards complete symmetry relative to the equator, i.e., the curve is approaching the zero line. Another option, which would imply that both curves are not accurate would be that the supercontinent is not moving laditudinally during the period of the superchron. In other words, both curves should be drawn

200

horizontally during the superchrons. If in fact the supercontinent is moving latitudinally during the superchron, then the inner core is able to maintain spherical symmetry within the outer core as all three core elements move toward the geocentric position. Based on the above logic, the next superchron will not occur until the next supercontinent forms.

Today's continents, as mentioned earlier are distributed fairly uniformly around the globe. If geophysicists were able to build a computer model tracking all continental movement along with their moment of inertia data (i.e., elevation including mountain ranges, volcanic activity, etc.), they might be able to estimate when the next geomagnetic reversal would occur. The last geomagnetic field reversal occurred about 780,000 years ago.

Finally, based on the GTME as it pertains to geomagnetism, the paleomagnetic secular variation would be controlled in the same way as geomagnetic reversals are. Currently, the north geomagnetic pole is drifting westward about 0.2 degrees per year and scientists are not able to provide an indisputable reason for this movement. Again the GTME attributes this movement to the spherical asymmetry of the inner core within the outer core and the asymmetry is a reaction to continental movement and is governed by the Law of Conservation of Angular Momentum. It is only when the inner core/outer core asymmetry becomes large that a geomagnetic excursion or geomagnetic reversal could occur.

The ability of the Earth's core elements to move away from their geocentric position, which scientists seem not to be aware of, is the foundation of the GTME.

References

DINOSAURS
The Dinosaur Heresies, 1986.......................Robert T. Bakker, PhD
A Field Guide to Dinosaurs, 2003.............Henry Gee & Luis V. Rey
The Illustrated Encyclopedia of Dinosaurs, 2006.....Dougal Dixon

EXTINCTION-Permian-Triassic
Extinction, 2006...Douglas H. Erwin
Rivers In Time, The Search For Clues to Earth's Mass Extinctions
    2000 ....Peter D. Ward
Evolutionary Catastrophes: The Science Of Mass Extinction
    1999....................................Vincent Courtillot

EXTINCTION-Phanerozoic
Catastrophes And Lesser Extinctions, 2004................Tony Hallam
Extinction..1987..Steven M. Stanley
Mass Extinctions And Their Aftermath..1997..............A. Hallam &
P.B. Wignall
Mass Extinctions And Sea-Level Changes....1999
Earth Science Reviews.............................A. Hallam & P.B. Wignall
What Caused The Mass Extinctions........................V. E. Courtillot
(*Scientific American* 263, no. 4:85-92 1990) *Time-calibration of
Triassic/Jurassic microfloralturnover,eastern North America
Technophysics.*1993.S.J. Fowell & P.E.Olsen & L.H. Tanner
Press/Pulse: A General Theory of Mass Extinction,
  GSA Conference paper...2006.................N.C. Arens & I.D. West
*Stability of Atmospheric CO2 levels across the Triassic/Jurassic
boundary. Nature* 6/7/01...Tanner, Hubert, Coffey & McInerney

EXTINCTION-Cretaceous-Tertiary

Extraterrestrial Cause For The Cretaceous-Tertiary Extinction
(Science, 6/6/80 Vol 208, No. 4448)
......Luis W. Alvarez, Walter Alvarez, Frank Asaro, Helen V. Michel
The Mistaken Extinction, Dinosaur Evolution and the Origin of
Birds, 1998 ....Lowell Dingus & Timothy Rowe
*T.rex And The Crater of Doom, 1997.... ....*Walter Alvarez
*Night Comes To The Cretaceous, 1998.......*James Lawrence Powell
*The Great Dinosaur Extinction Controversy, 1996..*Charles Officer
& Jake Page
Impact From The Deep-Scientific American, 9/18/06
..................Peter Douglas Ward
*K-T Boundary Issues..*Gerta Keller
(Science 264:641 1994)
End-Cretaceous Mass Extinction Event: Argument for terrestrial
*Causation(Science 238:1237-42....1987.................*Anthony Hallam
A Terminal Mesozoic Greenhouse: Lessons From The Past
*(Science 201:401-406 1978.................................*Dewey M. McLean
*Death Of The Dinosaurs....................................*Charles B. Officer
(New Scientist 2/20/93)
Dinosaur Experts Resist Meteor Extinction Idea, NY Times
10/29/85 ....Malcolm W. Browne
Rethinking What Caused the Last Mass Extinction, NY Times
11/6/07 ....John Noble Wilford
For Dinosaur Extinction Theory, A 'Smoking Gun', NY Times
2/7/91...John Noble Wilford
Paleoecological implications of Alaskan terrestrial vertebrate
fauna in latest Cretaceous time at high paleolatitude
(Geology 21 : 503-506 1993)......................Clemens & Nelms
Doubts On Dinosaurs: Yucatan impact crater may have occurred
before the dinosaurs went extinct. Scientific American 5/16/05
............................Barry E. DiGregorio
The End of The Cretaceous: Sharp Boundary or Gradual
*Transition? Science 1984.................*W. Alvarez, L.W. Alvarez, et al.

EXTINCTION/Ammonites, Nautilus

On Methuselah's Trail...1992............................Peter Douglas Ward
In Search Of Nautilus..1988............................Peter Douglas Ward
The Extinction Of The Ammonites....................Peter Douglas Ward
(*Scientific American* Vol 249 No. 4 October 1983)
*Late Maastrichtian And Earliest Danian Scaphitid Ammonites
From Central Europe: Taxonomy, Evolution And Extinction.*
(*ACTA Palaeontologica Polanica* 50 (4) 2005) ..Marcin Machalski

EXTINCTION-MISC.

Extinction Events Among Mesozoic Marine Reptiles...Nathalie
Bardet(*Historical Biology*, 1994, Vol 7. Pg313-324)
Phanerozoic Sea-Level Changes 1992....................Anthony Hallam
Study Solves Pangea Puzzle.................(*SCIENCEDAILY* 12/19/00
Atlas Of Mesozoic and Cenozoic Coastlines 1994
.... Alan G. Smith, David G. Smith, Brian M. Funnell
*Evidence for late Paleozoic and Mesozoic non-dipole fields
provide an explanation for the Pangea reconstruction problems*
.... Rob Van der Voo and Trond H. Torsvik
*Geometrical Approach to Torque Free Motion of a Rigid Body
Having Internal Energy Dissipation* ....
....Philippe L. Lamy and Joseph A. Burns
Earth-shaking 'wobble' may have killed dinosaurs
.... Thom Patterson
Plate tectonic maps-PALEOMAP PROJECT www.scotese.com
.... C. R. Scotese

*Environmental determinants of extinction selectivity in the fossil
record*...Shannon Peters

GEOMAGNETISM

Plate tectonics may control geomagnetic reversal frequency
....................................F. Petrelis, J. Besse and J.Valet
Early Pangea 'B' to Late Permian Pangea 'A'
.......................Muttoni,Kent,Garzanti,Brack,Abrahamsen,Gaetani

Index

acid-rain 5, 11, 25, 70, 91, 120
acritarchs 3
Actinoceratoidea 115
Actinofibrillae 106
Africa . . . . ix, 29, 52-53, 72, 79, 81-83, 105, 106, 111, 145
Agnathan 3
Agustinia 82
Alaska 24, 105, 203
albedo 5
Albian 127, 130
algae 12, 13
allometric 71
allosaurs 48
Allosaurus 41, 104
Alvarezes, Luis/Walter. 8-9, 11, 16-17, 22-24, 27, 33, 41, 104,
121, 203
Amargasaurus 81, 102
ammonites. 3-4, 22, 36, 66, 69-70, 114-135, 148, 153, 158, 204
amoeba 134
amphibians 4, 11, 24, 137-138
Amygdalodon 82
Ancyloceratida 127
Ancyloceratidae 129
Ancyloceratina 128
Anisoceratidae 129
ankylosaurids 105
anoxia 5, 87, 94, 110, 153
Antarctica 53, 79, 89, 111, 142
Antetonitrus 81
antipodal 151-152, 155, 161
Apatosaurus 48, 83
Appennine 7, 15
Archelon 112-113
archosaurs 137
Arens, Nan Crystal 6,202
Argentina 81-83
Argentinosaurus 58, 61-62

INDEX

Arthropleura 145
Asaro, Frank 11,203
Asia 72,79,81-82
asphyxiation, marine 94
Asteiriceratidae 129
asteroid 7, 9, 11, 18, 21, 27, 30-32, 40, 70-71, 86, 160
asthenosphere 29, 46
Atlantic . . . . ix, 4, 40, 89, 130, 137-138, 145, 159
atmosphere 5, 8-9, 42, 90, 93-94, 137
Australia 53, 138
avian/non-avian dinosaurs 41, 43, 49, 54-55, 57, 159
Bactritoidea 115
Badlands of Montana 31
Bakker, Dr. Robert T. 23, 99-100, 108-109, 202
Baltica 95
Baptistina 160
Barapasaurus 82
Bardet, Nathalie 112, 204
Barosaurus 48, 83
basalt . . . . . . xii, 3-5, 29, 40, 72, 88-89, 96, 100, 137, 148, 159-162
Batrachognathus 107
Bay of Biscay, Spain 109
belemnites 22, 66, 115, 158
benthic 3, 68, 118, 134, 150, 163
Berkeley 7, 9, 17, 24, 31, 121
Berriasian 114
Biarritz 121
Big Bend, Texas 107
biostratigraphy 135
biozonation 135
birds 11, 24, 49, 58, 70, 106-107, 114, 159, 203
bivalves 12, 109
Blikanasaurus 81, 102
Bochianitidae 129
bolide . . xi, xii, 5, 9, 11, 15, 20, 32, 40, 55-56, 73, 86, 100, 126, 137,
                                                              149, 160-161

INDEX

Bottacione, Italy 19
bottom-dwelling 3, 22, 66, 68-69, 121, 134, 150
brachiopods 2, 66
brachiosaurids 100, 102
Brachiosaurus 41, 48, 58
Brachytrachelopan 83
brontosaurs 60
Browne, Malcolm W. 32, 203
bryozoans 2
burrowing dinosaurs 25, 159
calcareous 13, 15, 20, 134, 158, 164
calcium carbonate 134-135, 158
Callovian 104
camarasaurids 100
Camarasaurus 48, 83
Cambrian 134
Camelotia 81
camerae 117, 124, 134
cameral fluid 124, 126, 133
CAMP 4, 137-139, 148, 160, 162-164, 188
Campanian 105, 112
camptosaurids 100
Canada 79, 105, 111, 138
Carbon Cycle 33
carbon isotope C12 92
carbon isotope C13 92
carbon dioxide 5, 90-91, 94
carbon isotope 92
carbonate, see calcium carbonate . . . .
Carboniferous 39, 134, 145, 147, 149, 160, 172-174
Carey, Samuel Warren 76
Carinodens belgicus 111
Cedarosaurus 81
Cenomanian 61, 110- 111
Cenomanian/Turonian 110-111
Cenozoic 25, 86, 99, 108, 115, 158, 160, 204

Cephalopoda 115
cephalopods 116, 133
ceratopsids 100
Ceratosaurus 104
cetiosaurids 100
Cetiosaurus 82
CFBs 88
champsosaurs 24, 31
cheloniids 109
Chicxulub . . . . xii, 5, 17, 40, 56, 71, 100, 125, 160
China 62, 79, 82, 95-96, 104-105, 148, 159, 162
chron 8, 162
Chubutisaurus 82
Cimmerian Plate 146-148, 162-164
cladogram 48
clams 109, 158
clay layer 7-8, 12, 15-19, 20, 24, 27, 29-31, 57, 100, 121, 126,131
Clemens, Dr. William A. 24, 31, 121, 203
Clidastes 112
climate 2-3, 24, 49, 91, 137
coccoliths 5, 13, 20
Coelophysis 48
Coffey 202
Coleoidea 115, 119
Colorado 83, 103
comet 32, 86
conch 130
conifers 47
Conservation of Angular Momentum 95, 142, 157,182,192
Conservation of Rotational Kinetic Energy 179
Contessa, Italy 15
Continental Drift Theory 167
continental plate symmetry 192
CoPlot iv
Core Elements of Earth 173
core shift . . . . xii, 36-40, 76-80, 84, 95-98, 138-144,

148, 151, 157-164

core/mantle 29, 40, 88, 97, 139

corkscrew shaped shell 116, 119

couplets 6, 127, 138, 148, 150, 152-153, 155, 161-162

Courtillot, Vincent 33, 202

crabs 118

Cretaceous .. xii, 4-5, 7, 11, 13, 15, 23-25, 27, 31-32, 41-42, 48-49,
52, 54, 59-66, 68-71, 86, 96, 99, 102,
104-135, 143, 145, 153, 157-161, 164,203

Cretaceous Normal Superchron 187,192

Cretaceous/Cenozoic 160

Cretaceous/Paleogene 7

Cretaceous/Tertiary (see K-T) .... xii, 4, 11, 42, 100, 203

Croc 58

Crocket 19, 30

crocodiles 11, 24, 31-32, 70

crocodilians 58, 63, 159

crurotarsans 196

cryptoclidids 109

Crystospermaceae 138

Ctenochasma 107

cuttlefish 115

Cymbospondylus 114

Danian 4,115,124,130,158,204

Datousaurus 82

Deccan Traps ... xii, 5, 28-30, 33, 40, 71-72, 88-89, 159-160,190

delta 13C 92-95,97-98

Denmark 8, 18, 30

Deshayesitidae 129

Devonian 2, 134, 149-150, 153, 163

Dicraeosaurus 83

DiGregorio, Barry E. 203

Dingus, Lowell 203

dinoflagellates 66

dinosaur/mammal 75, 85

dinosaurs .... xi-xii, 4, 23-25, 31-33, 35, 37, 39-43, 46-49, 51-55,

57-60, 62-63, 70-76, 79, 85, 95, 100, 103, 105-106, 119, 137, 142, 158-159, 202-203

diplodocids 100, 102
Diplodocus 41, 48, 83
Diplomoceratidae 129
dipole 144
displaced magnetic axial dipole model 174
Dixon, Dougal 202
dolphins 114
Douvilleiceratidae 129
downsizing 43, 51, 54, 59, 61, 64, 68, 100
dragonflies 145
durophagous 118, 123
Edmontonia 105
eggs 33, 48-49, 62-63, 139, 159
Egypt 62, 115
ejecta 9, 20
elasmosaurs 4, 109
Elasmosaurus 64
elephant 72
end-Triassic 3, 5, 37, 137, 143, 157, 164
Endoceratoidea 115
end-Devonian 149, 163
end-Permian 85
Eobrontosaurus 83
Eocene 134
Eoraptor 48
epicontinental seas 87, 152
equator 79-80, 84, 110, 149, 163
equatorial 37, 76, 142-143, 146, 157, 163
Erketu ellisoni 83
Erwin, Douglas H. 202
Ethiopia 89
eugubina 12
euoplocephalids 100
Euoplocephalus 105

INDEX

Eurasian Plate 29
Europe 79, 81-83, 109, 111-112, 125-126, 138, 145, 204
Euskelosaurus 81
eustatic 2, 86, 150, 152-153, 161
extinction.... see mass extinction . . . .
extinction of ammonites. 3, 4, 36, 66, 69, 70, 115, 119-134,148,
153, 158, 204

extinction of dinosaurs . . . . . xi-xii, 4, 23-25, 31-33, 41-43, 48-54,
59-63, 70, 75, 100, 119, 142, 159, 203
extinction of forams . . . . see forams
extinction of inoceramids 109-111, 158
extraterrestrial . . . . xi, 11, 27, 30, 35, 42, 49, 86, 203
Exxon Vail Curve 86, 153-154
FBV 40
ferns 138
flood basalt volcanism 88-91,188,197
forams 4, 12-13, 15, 20, 25, 36, 66-68, 134, 135, 142, 158
Fort Peck, Montana 24, 31
Fowell, S. J. 138, 202
fungal spike 91
Funnell, Brian M. 204
fusulinids 134
Gallodactylus 107
gap, ammonite at Zumaya 122, 130
gastropods 66
Geocentric Axial Dipole (GAD) 169
geomagnetic field reversals 192,198-201
gigantism. .xii, 36, 39-40, 51, 57-58, 60-61, 63, 67, 69-70, 73, 75-76
85, 95, 142, 157
Giraffatitan 83
giraffe 60
glaciation 5, 86, 149
glaciers 2, 86
Globigerina eugubina 12
Globotruncana 12

INDEX

Glossopteridaceae 138
glossopteris 91,168
Gnathosaurus 107
Gobi 83
Gondwana 52-53, 59, 145, 149, 150, 163
graptolites 2
gravitation .... xii, 6, 13, 20,35-77, 80, 84-85, 95-98, 122-127,
                    130-135, 141-150, 152, 155-162, 164
gravitational gradient 36, 95, 142, 150, 160
gravitational increase 48, 66, 120, 123-127, 133, 138, 148,
                    152, 157-158, 161
gravity ... I, iii, vii, viii, x-xii, 5, 6, 13, 19, 35-39, 44-46, 48, 54, 57,
                64, 71-73, 75, 76, 79, 85, 95, 98, 107, 111, 114, 119,
                130, 133, 138, 139, 141-143, 145, 146, 150-153, 155,
                157, 159, 161-165
gray/white limestone 15-17, 20, 29
Greenland 79
Gubbio, Italy 7-8, 12, 14, 17-19, 29-30
hadrosaurs 24, 100
Hainosaurus 111
Halisaurus 112
Hallam, et al. sea level curve 196
Hallam, Anthony 138,149,153,202-204
Hamitidae 129
Hamulinidae 129
Hangenberg Event 3, 149
Hawaiian Islands 20, 30, 90
hematite 7
Hemihoplitidae 129
Hendaye 121
henodontids 109
herbivores 22, 100, 103, 105
Herrerasaurus 48
hesperornids 109
Heteroceratidae 129
heteromorph 123-124, 126-127, 133

212

INDEX

Himalayas 29
Hokkaido 110
hoploscaphites 125-126, 130
Hotton, Carol 31
Huayangosaurus 104
Hutchinson, Dr. Howard 31
hydrogen sulfide 5, 94, 97
hyponome 117
hypsilophodontids 100
Iceland 33
inclination angle of Earth's magnetic field 169
ichthyornids 109
ichthyosaurs 4, 64-66, 111, 114, 159
igneous provinces 40, 88, 159
iguanodon 100, 106
impact .... xi-xii, 5-7, 9, 11, 15, 16-33, 40, 49, 52, 55-56, 68, 70-
            71, 86, 100, 120-121, 124-126, 133, 137, 149, 160-
            161, 203
implosion depth 118, 122, 124, 126
India 28-29, 53, 71-72, 79, 82, 88, 145
inner core. 37, 46, 76, 79, 84, 95-98, 159,162
inoceramids 109-111, 158
insects 3, 22
iridium 8-9, 15, 18-21, 24, 27, 30, 70-71, 86, 126, 149
Isanosaurus 81
isotope 9, 92, 135, 137
jawless fish 3
Jeholopterus 107
Jingshanosaurus 82
Jurassic 3-4, 37, 41, 48, 60, 65, 69, 80, 99-100, 104-105, 107,
            108, 114-115, 130, 137-138, 143, 145, 148, 157-158,
            160, 162, 164, 202
Jurassic/Cretaceous 100, 104-105, 107
K/T   See K-T ....
Karoo 89
Keller, Gerta 25, 203

Kellwasser Event 3, 149
Kiaman Reverse Superchron 187,192
Kimmeridgian 104, 107
Koshland, Dr. Daniel E. 33
Kotasaurus 82
Krakatoa 9, 21
Kunmingosaurus 82
K-T .... viii, xi, xii, 7, 9, 11-14, 19-21, 23-25, 27-33, 37, 40, 86, 88,
         96, 99, 100, 108, 110, 115, 118-126, 130, 131, 133,
                      135, 142, 158, 160, 161, 164, 190,203
Laki 33
Landman, Dr. 126
latitude 2, 24, 79-80, 84, 135, 149-150
latitudinal 95, 138, 142, 150-151, 157, 160
Laurasia.............................. 52-53, 59, 145, 148
Laurasia/Gondwana 52, 59, 145
lava............................. 7, 28-29, 72, 90, 96, 138
Lesotho, Africa 81-82
Lessemsaurus 81
Lilliput Effect 12, 68, 135
limestone........................... 7, 12, 15, 17, 20, 29, 112
Lindgren 112
LIPs 88
lithosphere 71, 87, 155
live birth 65, 114, 139, 159, 164
lizards 24, 58, 66, 70, 111
longitudinal 95, 151, 160
Lufeng, China 82
Lusotitan 83
Maastrichtian ... 68, 105, 110-112, 115, 120, 122, 125-127, 161, 204
Machalski, Marcin 204
Madagascar.............................. 53, 72, 82, 89
magma ix
magmatic province 4, 137
magnetostratigraphy 7-8
Malawi, South Africa 82

INDEX

Malawisaurus 82
Mamenchisaurus 82
mammals 11, 43, 49, 54, 57-58, 60, 62-63, 75, 85, 114, 159
Manasquan River Basin 126, 131
mangrove 62
Manicouagan Crater 137
mantle 19, 29, 40, 46, 87-90, 96-98, 139, 148, 162
Marshall, Charles 109
marsupials 62, 159
Martinique 90
mass extinction.................. x-xii, 1-2, 5-6, 13, 20, 27, 32,
          35-40, 55, 73, 100, 108-109, 119, 126, 137, 145, 149,
          150, 153,157, 161-163, 165, 186-188,191,195-196
Massospondylus....................................... 82
McInerney... 202
McKee, Chris 9
McLean, Dewey M. 32-33, 203
McLeod, Kenneth G.. 110
Mediterranean ammonites 120
Melanorosaurus 81, 102
Mesozoic .... xii, 7, 24, 35-36, 39-43, 47-49, 58-59, 62, 73, 75-76,
          79, 84-86, 99-100, 106, 108-109, 115, 118-119, 122-
          124, 127, 130, 132-134, 143, 149, 152-153, 155, 157-
                                        162, 203-204
metacarpals 107
meteor .... xii, 8-9, 32, 42, 203
meteorites 8, 9
methane 5, 37, 91-92, 94-95, 97-98, 130, 145-146, 162-163
methyl hydrates. . 37, 91, 93, 94-95, 97, 98, 145-146, 157, 162, 164
metriorhynchids..................................... 109
Meyaneuropsis permiana 145
Michel, Helen V. 11, 203
Miocene... 134
mid-Cretaceous 54, 59, 106, 158-159
Milankovich cycles 152
millipedes 145

INDEX

Mississippian 115
mixosaurids 109
Mixosaurus 114
mollusks 115
moment of inertia of the Earth 180
Mongolia 82-83
Montana 24-25, 31, 83
mosasaurids 109
mosasaurs 4, 36, 65- 66, 111-112, 159
Mount Pelee, Martinique 90
Mount Pinatubo, Philippines 90
Mount St. Helens, USA 29, 90
nanoplankton 15
NATIP 159
Nautiloidea 115, 118, 133
nautilus 69, 115-119, 122-123, 126, 133-134, 204
Nelms 203
Netherlands 112
New Zealand 8, 53,111
Newark 89
Newell, Norman 152
Newton, Isaac 35-36, 39, 44-45, 52, 75, 77, 142
Niger 81
Nigersaurus 81, 102
nodosaurids 105
Norian 4, 102
Norway 143
Nostoceratidae 124, 129
nothosaurids 109
nutation damping 183
octopus 69, 115, 119-120, 133
octupole 172
Officer, Charles 203
Ojo Alamo Sandstone 24
Oklahoma 81
Olsen, P. E. 138, 202

INDEX

Omeisaurus 82
Ordovician 2, 150, 153, 163
ornithocheirids 109
ornithodesmids 109
Orthoceratoidea 115
Oryctodromeus 25
oscillations of inner core 98
outgassing 33, 90
Oviraptosaurs 41
Oxfordian 104, 107
oxygen 5, 87, 94, 97, 109-110, 135
ozone 91
pachycephalosaurids 100
Pachycephalosaurus 41
pachypleurosaurids 109
Pacific Plate 90
Page, Jake 203
Paleocene flora 24
paleoequator 146, 148, 162, 164
Paleogene 7, 124-125, 134
paleolatitude 90,xxx
paleomagnetism 167
paleomagnetists 167-172
Paleotethys, see Tethys . . . .
Paleozoic 86, 146
palynofloral 138
palynomorph 138
Pangea A & B 170
Pangaea 42
Pangea 3-5, 35-37, 39-40, 42, 46-48, 51-53, 57, 59, 63,
                    67, 72, 75-76, 79-80, 84-85, 87, 91, 95-98,
                       138, 142-146,148, 151-152, 155, 157-158,
                                          160-164, 204
panoplosaurids 100
Panthalassa Ocean 178
Parahoplitidae 129

Paralititan stromeri 59, 62
Patagosaurus 83
Peltaspermaceae 138
Permian . xii, 3, 5, 37, 39, 40, 85, 91, 93, 95-97, 124, 130, 134, 142,
                153, 157, 160, 162, 163, 168,171-174.186,202
Peters, Shanon 152,204
photosynthesis 9, 22, 120
phragmocone 117
Phuwiangosaurus 82
Pinacosaurus 105
pink limestone 7, 15, 20, 29
Piton de la Fournaise 29
placental 62
placocheliids 109
placoderms 3
planktic 4, 134-135, 142, 158
planktonic 12, 66
plants 3, 11, 22, 25, 42
plate, tectonic ..... ix-xii, 29, 39, 40, 42, 51, 56, 71-72, 76, 87, 90,
                95-96, 98, 138, 143-144, 146-148, 152, 155,
                                157-160, 162-164
plateaus 155
Plateosaurus 81, 102
Plate/Reversal Hypothesis 192-197
Platypterygius 65
plesiosaurids 109
plesiosaurs 36, 63-64, 114
pliosaurids 109
pliosaurs 112
plume 29, 72, 87-90, 96-98, 139, 148, 162
plutonium 9
polacanthids 105
polar 24, 87, 135, 149, 152, 162
polarity, magnetic 7-8
pole 8, 46, 110, 149-150, 162-163
pollen 24, 31, 42, 138

INDEX

pollution 70
polycotylids 109
Portugal 82-83
Powell, James Lawrence 203
ppb 20
Press/Pulse 5-6,202
prosauropods 80-81, 102
protoceratopsids 100
protostegid turtles 32, 109, 112
protozoa 67
Prudhoe Bay, Alaska . . . . ·24
Pterodactyloidea 107
pterodactyls 48
Pterodactylus 48, 107
pterosaurs 48, 100, 106-107
Ptychoceratidae 129
Pyranees 120
pyroclastic 30, 90
quadrapole 172
Quetzalcoatlus northropi 107
radiolaria 66
Rajmahal 89
Rapetosaurus 82
Raup, David 55
rarefaction analysis 110
regression 2-3, 5, 31, 86-87, 95, 122, 152, 155, 157, 161
regression/transgression 86-87, 155
regressive/transgressive couplets . . 6, 127, 138, 148, 150, 152-153
155, 161-162
Repenomamus giganticus 62, 159
Repenomamus robustus 62, 159
Reunion Island 29-30
reversal of poles 8, 162
RGG 54-59, 61-63, 67, 69, 85
Rhaetian 4
rhamphorhynchids 109

Rhamphorhynchoid 107
Rhamphorhynchoidea 107
Rhamphorhynchus 48, 107
rhomaleosaurids 107
ribbed shells, ammonites 118, 123, 125-127, 130
Rigby, Dr. J. Keith, Jr. 24
Riojasaurus 81
Rocchia, Robert 19-20, 30
rotalinids 134
Rowe, Timothy 203
salamanders 11
Saltasaurus 82
Sarcosuchus imperator 58-59
sauropods 36-37, 48-49, 52, 54, 57, 58, 60-61, 72, 79-81,
                              84, 100, 102-103, 119, 142, 145, 157-158
Sauroposeidon 58, 81, 102
Scaglia rossa 7
scaphitid 125-126,204
Scelidosaurus 104
Scotese, Christopher 192
sea level 151-155,192-193
Seismosaurus 83
Serra Geral 89
Seychelles 53
shastasaurids 109
Shunosaurus 82
Siberia 79, 138
Siberian Traps . . . . xii, 3, 40, 87-91, 94-95, 97-98, 159,188
Signor, Philip 119, 122
siphuncle 117, 122-124, 126, 133-134
Siverson 112
Smith, Alan G. & David G. 204
snail 69
snakes 11, 24, 58, 66, 70, 111
soils 137
Sordes 107

south China Blocks 95-96, 148, 162
Spain 83, 109, 119-120
spinosaurs 100, 106
Spinosaurus 106
squid 69, 115, 119-120, 133
Stanley, Steven M. 122, 149, 167
stegosaurians 104-105
stegosaurs 48, 100, 103
Stegosaurus 41, 104
stenopterygiids 109
stepwise extinctions 59
Stevns Klint 8, 18-19, 30, 125
sulfur dioxide 33, 137
Sumatra 9
superchrons 187,201
super-continent 35-36, 42, 46-48, 51-53, 57, 59, 67, 72,
75-76, 85, 95, 142, 144, 146, 149, 151, 163
supernova 9, 21
Supersaurus 58, 83
sutures, ammonite 118-119
Switzerland 81
Takashi, Akinori 110
Tanner, Lawrence H. 137, 167
Talarurus 105
Tanzania 83
tectonic .. ix-xii, 29, 39, 40, 42, 51, 56, 71, 72, 76, 87, 96, 98, 138,
143-144, 146, 152, 155, 157-160, 162-163
tektites 149
teleosaurids 109
Tenuipteria 110
Tertiary .... xii, 4, 7, 11, 25, 42, 68, 86, 100, 120, 130, 159, 203
Tethys 106, 109, 145-146, 162
tetrapod 3, 99-100, 109
Texas 107
textularinids 134
Thailand 81-82

therapsids 4, 137, 139
thermoregulation 104, 106
theropod 48-49, 104, 106
titanosaurs 60-62, 100, 102-103
Tithonian 104
toxocheliids 109
trackways 61

transgression 2, 3, 5, 6, 86-87, 95, 122, 127, 130-131, 133,
152,155
transgression/regression 86
transgressive 6, 122, 127, 138, 148, 150, 152-153, 155, 161-162
Triassic . . . . . . xii, 3-5, 37, 41, 47-48, 65, 85, 93-97, 102, 108, 114,
116, 124, 130, 137-138, 143, 145, 148, 153, 157, 160,
162, 164, 167
Triassic-Jurassic 3, 37, 130, 145, 148, 157, 162, 164, 167,195
tribolites 2
triceratops 31, 41, 100
Trochleiceratidae 129
tsunamis 70-71
tubercles 124-126, 130
tuberculation 118, 123, 125-127
Turonian 110-111
Turrilitidae 124, 129
turtles 11, 24, 32, 58, 63, 112, 114
Tycho 160
tyrannosaurs 100
Tyrannosaurus rex 106
ultraviolet radiation 91
Uniformitarianism 29
Utatsusaurus 114
Valanginian 114
Vesuvius 90
viviparous 65-66, 111, 164
volcanism . xi, xii, 3-6, 9, 20, 28-30, 32, 33, 40, 55, 71, 73, 87-91,
94-98, 100, 137, 139, 159-161, 164

Vulcanodon 82
Ward, Peter Douglas. . . .109, 119, 120-124, 126, 130-131, 133, 139,
                                                    202-204
Wegener Alfred . . . . ix-x, 76,167
Whale 71
wide-gauge trackways 61
Wiedmann, Jost 120-121
Wignall, P. B. 138, 149, 153, 167
Wilford, John Noble 31, 126, 203
wobble of Earth 40, 160,190
Yucatan 203
Yunnan, China 82
Yunnanosaurus 82
Zimbabwe, Africa 81, 82
zooplankton 120
Zumaya, Spain 109, 119-121, 130

www.ingramcontent.com/pod-product-compliance
Lightning Source LLC
Chambersburg PA
CBHW070417290526
45791CB00005B/1728